Green Energy and Technology

Climate change, environmental impact and the limited natural resources urge scientific research and novel technical solutions. The monograph series Green Energy and Technology serves as a publishing platform for scientific and technological approaches to "green"—i.e. environmentally friendly and sustainable—technologies. While a focus lies on energy and power supply, it also covers "green" solutions in industrial engineering and engineering design. Green Energy and Technology addresses researchers, advanced students, technical consultants as well as decision makers in industries and politics. Hence, the level of presentation spans from instructional to highly technical.

Indexed in Scopus.

Indexed in Ei Compendex.

More information about this series at https://link.springer.com/bookseries/8059

Kailong Liu · Yujie Wang · Xin Lai

Data Science-Based Full-Lifespan Management of Lithium-Ion Battery

Manufacturing, Operation and Reutilization

Kailong Liu
Warwick Manufacturing Group (WMG)
University of Warwick
Coventry, UK

Yujie Wang
Department of Automation
University of Science and Technology
of China
Hefei, China

Xin Lai
School of Mechanical Engineering
University of Shanghai for Science
and Technology
Shanghai, China

ISSN 1865-3529   ISSN 1865-3537 (electronic)
Green Energy and Technology
ISBN 978-3-031-01342-3   ISBN 978-3-031-01340-9 (eBook)
https://doi.org/10.1007/978-3-031-01340-9

This Springer imprint is published by the registered company Springer Nature Switzerland AG
The registered company address is: Gewerbestrasse 11, 6330 Cham, Switzerland

Foreword by Prof. Qing-Long Han

Advanced technologies that accelerate the delivery of renewable and sustainable energy systems will contribute to low-carbon applications and environmental pollution reduction. A battery is a primary energy storage source and a key part of the energy transition, where a lithium-ion (Li-ion) battery is particularly promising due to its excellent performance such as high energy density, low self-discharging rate, and continually decreasing price. According to the market forecast, the global Li-ion battery market size would reach US$105.6 billion by 2026, making it become the key technology to transportation electrification and provide stationary energy storage for electrical grids.

However, a battery will inevitably age with time, while its performance is a consequence of multiple coupled mechanisms affected by various elements such as battery manufacturing, operation, and environmental condition. Specifically, the manufacturing process will first directly determine the battery initial performance. Then, the battery with high performance will be widely operated in electric vehicle applications until the capacity reaches 80% of its nominal value. After that, the retired battery will be reutilized in second-life applications such as grid energy storage to make full use of its residual value. To further benefit the environment and save cost, the non-function battery will be recycled finally. During such full-lifespan of battery, various cases such as overcharging/discharging, internal/external short circuits could damage battery efficiency, and even cause safety issues like a thermal runaway, combustion, and explosion. In this context, effective battery management becomes necessary to ensure its safety and efficiency during the full-lifespan.

The recent rapid developments of automation science and big data analytics provide data science-based solutions to manage batteries. The book attempts to capture the latest data science technological advancements in battery management and apply these to battery full-lifespan cycle including three main stages: manufacturing, operation, and reutilization. It covers the research results of Dr. Kailong Liu, Dr. Yujie Wang, and Dr. Xin Lai in data science-based techniques for battery full-lifespan management. This book starts with the background and motivation of the Li-ion battery and its management. Then the vision of the battery full-lifespan management and its each stage are provided. It continues with the descriptions of several key management tasks within the manufacturing, operation, and reutilization stages of battery full-lifespan, while various classical and emerging data science-based strategies to handle these tasks are also illustrated with the well-proven case studies. Finally, the key challenges, future trends, and promising data science technologies to further improve this research field are discussed. Thus, this book focuses on the intersection of battery data science engineering and offers a systematic background and technology information on the data science-based management for full-lifespan of the Li-ion battery.

Dr. Liu's research experience lies at the intersection of AI and electrochemical energy storage applications, especially data science and automation engineering in battery manufacturing and operation management. Throughout over 10 years of research experience, he has focused on the discovery of new knowledge and understanding with a focus on application-based research and academic research with industrial impact. Currently, he leads the data science activities to benefit battery management in WMG, where he works closely with many academic institutions (e.g. Oxford, UCL) and battery or automotive industries (e.g. Jaguar, Aston Martin, UKBIC, Britishvolt, and Varta Storage). The impact of his AI and data science research has been significant, as key elements have been integrated into the practice of a number of industrial partners. Dr. Wang has been with the School of Information Science and Technology, University of Science and Technology of China, where he is currently an Associate Professor at the Department of Automation. He is currently the Secretary of System Simulation Committee of Chinese Association of Automation, and the Director of IEEE PES Electric Vehicle Technical Committee. His current research interests are data analysis of renewable energy systems and data science management of lithium-ion batteries. Dr. Lai's research interests include battery fault diagnosis, secondary utilization of the retired batteries, and recycling and remanufacturing of the waste batteries. In recent years, he is committed to applying data science and AI to these research fields. This book represents the persistent efforts and research experiences of them and their groups in developing core data science strategies for battery full-lifespan management.

I highly recommend this book to researchers/Ph.D. students and engineers who are working in the fields of battery management and data science. First, this book has a solid technical content and can provide the basics of battery manufacturing, operation, and reutilization. With various practical case studies, readers could also understand how to derive data science-based strategies to achieve efficient management of these three important stages during battery full-lifespan. Second, this book covers the hot and promising research topics from Dr. Liu, Dr. Wang, and Dr. Lai, who are the rising star in this field. They all have a rich research experience and achieve excellent results in promoting advanced data science and AI-based strategies for battery full-lifespan management. In summary, considering the rapid development of data science-based battery management technologies, this book is significantly important for AI and battery applications, which can be extremely attractive to readers in the community.

Qing-Long Han
Pro Vice-Chancellor and Distinguished Professor
Member of the Academia Europaea
IEEE Fellow, FIEAust Fellow
Swinburne University of Technology
Melbourne, Australia

Foreword by Prof. Jinyue Yan

The energy transition is one of the most complex challenges facing societies today, requiring widespread social and technological changes over several decades. The increased penetration of renewable energy and the responses to the flexibility of energy systems, energy storage technologies including batteries will play a significant role in the future energy transition. However, a battery inevitably degrades with time, losing the capacity to store charge and deliver it efficiently. Battery management systems are significant for battery performance including efficiency, capacity, and safety.

Meanwhile, the rising dynamic pressure on variable renewable energy penetration and the fluctuation of energy demand requires that we better understand the energy nexus, i.e. connections between energy and other sectors by using interdisciplinary approaches. This book is one kind of such efforts to apply data science for the management of batteries.

With the rapid advancements of "Big Data" analytics and AI, data science is becoming one of the promising tools to respond to the challenge of battery management due to its superiority of being flexible. The full-lifespan of a battery consists of three crucial stages: manufacturing, operation, and reutilization. A manufacturing stage is to produce and determine battery initial performance. In an operation stage, battery capacity degrades from 100 to 80% to supply/absorb electrical power for high-demand applications (e.g. electric vehicles). In a reutilization stage, the battery capacity degrades below 80% for the reduced demand applications (e.g. grid energy storage) and then is recycled. As all these three stages are significantly correlated to battery performance, a systematic exploration simultaneously covering battery full-lifespan (manufacturing, operation, and reutilization) management is urgently required.

This book is an academic monograph covering full-lifespan management of Li-ion batteries by using data science approach. It is also a collected achievement of scientific research by the authors. The book starts with the introduction of the Li-ion battery's role in the energy transition. Then the battery full-lifespan management and key challenges are showcased. Afterwards, the efforts of designing suitable data science solutions to handle each challenge in the management of battery manufacturing, operation, and reutilization are presented with detailed case studies. Finally, future development trends are prospected and discussed.

I highly recommend this book to the academics, engineers, and practitioners in both applied energy and AI communities. The book will provide readers with the knowledge of battery management over lifespan including manufacturing, operation, and reutilization with the integration of technological innovations and engineering practices by means of data science. The step-by-step guidance, systematic introduction, and case studies to the topic enable the book useful to audiences of different levels, from graduates to experienced engineers.

Stockholm, Sweden   Prof. Jinyue Yan
Spring 2022

Preface

To achieve net-zero carbon emission targets around the world, numerous works regarding green energy have been explored, offering plenty of potential sustainable and low-carbon technologies. Owing to the superiority in terms of high energy density and reliable service life, lithium-ion (Li-ion) battery represents one promising and efficient energy storage solution for many applications such as electrical devices (i.e. mobile phone, laptop), transportation electrification (i.e. electrical vehicle, electrical ship, electrical plane), and smart grid. However, due to the frequently occurring electrochemical reactions, after battery is manufactured, it will inevitably age with time, losing the capacity to store charge and deliver it efficiently. Besides, the safety, efficiency and performance of a battery would be also directly affected by different operating cases, making related management solutions become crucial and necessary.

A hologram of battery full-lifespan consists of three main stages including manufacturing, operation (battery capacity from 100 to 80%) and reutilization (battery capacity below 80%). To ensure the safety, efficiency, and performance of battery during its full-lifespan, battery management solutions are developing rapidly in the international market and become a hot research topic. However, there are still numerous technical challenges in battery full-lifespan applications, demanding the development of state-of-the-art technologies. Firstly, battery properties such as cost, reliability, energy density, and life are directly determined by its manufacturing process. It is thus vital to develop suitable solutions for understanding and analysing battery intermediate manufacturing processes in the pursuit of smarter battery manufacturing management. Besides, owing to the battery's complicated electrochemical dynamics, numerous tasks such as battery state estimation, ageing prognostics, fault diagnosis, and charging management must be done to well and efficiently manage battery during its operation stage. Moreover, to make full use of battery, the retired battery from electric vehicles with a capacity below 80% needs to be reutilized in second-life applications such as grid energy storage. Furthermore, to comply with environmental and health benefits, batteries need to be recycled when they get either spoilt or non-functional.

With the rapid development of AI and machine learning, data science-based applications have drawn much attention and become a research hotspot in the field of battery full-lifespan management. After well designing proper data science solutions, significant enhancement can be achieved for more effective battery management from the aforementioned three stages. However, as relatively new and prospective research, currently there is no book to systematically introduce and describe the battery full-lifespan management particular from the data science application perspective to our best knowledge.

This book comprehensively consolidates studies in the rapidly emerging field of battery management. The primary focus is to overview the new and emerging data science technologies for full-lifespan management of Li-ion batteries, which are categorized into three groups, namely (i) battery manufacturing management, (ii) battery operation management, and (iii) battery reutilization management. The book is broken down into 7 chapters which contain illustrations and tables. Several key features are described in this book as:

- The concept of battery full-lifespan management is proposed, and the related data science solutions to handle its key tasks are described.
- A systematic introduction of the expanding data science technologies in battery full-lifespan management is provided.
- Data science-based management of battery manufacturing including electrode manufacturing and cell manufacturing is introduced with case studies.
- Data science-based management of battery operation including operation modelling, state estimation, ageing prognostics, fault diagnosis and battery charging is introduced with case studies
- Data science-based management of battery reutilization including echelon utilization and material recycling is introduced with case studies.
- Data science is proven to be a promising route to improve battery full-lifespan management.
- The key challenges, future trends, and promising data science technologies to further improve battery full-lifespan management are discussed.

This book will attract the attention of academics, scientists, engineers, and practitioners. It will also be useful as a reference book for graduates and PhDs working in the related fields. Specifically, the audience could get the basics of battery manufacturing, operation and reutilization, and also the information on related data science solutions. We hope the book will inform insights into the feasible, advanced data science design for battery manufacturing, operation, and reutilization management perspectives, further boosting the development of reliable data-based modeling, analysis, and optimization technologies to enhance battery performance.

Coventry, UK   Kailong Liu

Acknowledgments

The authors also would like to thank (i) Prof. Kang Li (University of Leeds, UK), Prof. Zonghai Chen (University of Science and Technology of China), Prof. Zhongbao Wei (Beijing Institute of Technology, China) for their constructive feedback on this monograph; (ii) a number of professionals and group leaders who collaborate with us to generate the resource of this monograph including Prof. James Marco (University of Warwick, UK), Prof. Dhammika Widanage (University of Warwick, UK), Prof. Anup Barai (University of Warwick, UK), Prof. Aoife M. Foley (Queen's University Belfast, UK), Prof. Harry Hoster (Lancaster University, UK), Prof. Remus Teodorescu (Aalborg University, Denmark), Prof. Josep M. Guerrero (Aalborg University, Denmark), Prof. Van Mierlo Joeri (Vrije Universiteit Brussel, Belgium), Prof. Furong Gao (Hong Kong University of Science and Technology, Hong Kong Special Administrative Region), Prof. Minggao Ouyang (Tsinghua University, China), Prof. Chenghui Zhang (Shandong University, China), Prof. Xiaosong Hu (Chongqing University, China), Prof. Yunlong Shang (Shandong University, China), Prof. Yuejiu Zheng (University of Shanghai for Science and Technology); (iii) funding support from the National Key R&D Program of China under Grant No. 2020YFB1712400, the UK High Value Manufacturing Catapult Project under Grant No. 8248 CORE, and the University Synergy Innovation Program of Anhui Province under Grant No. GXXT-2019-002; and (iv) Mr. Anthony Doyle, Executive Editor, Engineering, Springer, 236 Gray's Inn Road, Floor 6, London WC1X 8HL, UK, for his encouragement to write this monograph.

Contents

About the Authors

Dr. Kailong Liu is currently an Assistant Professor with the Warwick Manufacturing Group (WMG), University of Warwick. He received the Ph.D. degree from the Energy, Power and Intelligent Control group, Queen's University Belfast (QUB), UK, in 2018. He was a Visiting Student Researcher at the Tsinghua University, China, in 2016. His research interests include data science and AI techniques, with applications to battery manufacturing and management; the development of new and advanced control and optimization technologies for energy management in electric vehicles and battery systems, with the aim to improve their efficiency, safety, and sustainability. He has produced more than 50 peer-reviewed Q1 journal, transactions papers and authored five patents related to battery manufacturing and management, with a total citation of over 2300, and an H-index of 25. He is currently the Editorial Board Member of *Renewable and Sustainable Energy Reviews* (IF: 14.98), *IEEE/CAA Journal of Automatica Sinica* (IF: 6.17), *Frontiers in Energy Research* (IF: 4.01), *Control Engineering Practice* (IF: 3.48), and *Advances in Applied Energy*. He has been served as the Guest Editor for several international journals related to AI and battery management (*Energy, International Journal of Electrical Power and Energy Systems, IEEE Transactions on Transportation Electrification, IEEE JESTPE, IEEE JESTIE, Transactions of the Institute of Measurement and Control*, etc.), the Section Chair or Program Committee Member for many international conferences (IJCNN2021, SCCI2021, EMET2021, WCCI2020, CEC2020, AIAM2020, etc.), and the Outstanding or Active Reviewer for over 30 international journals in the research field.

Dr. Yujie Wang received the Ph.D. degree in Control Science and Engineering from the University of Science and Technology of China, China, in 2017. Since 2017, he has been with the School of Information Science and Technology, University of Science and Technology of China, where he is currently an Associate Professor at the Department of Automation. He is currently the Secretary of System Simulation Committee of Chinese Association of Automation, the Member of Simulation Technology and Applications Committee of China Simulation Federation, and the Director of IEEE PES Electric Vehicle Technical Committee. He has received the

Excellent Doctoral Dissertation of the Chinese Academy of Sciences in 2019, the Excellent Paper Award of the 34th World Electric Vehicle Symposium and Exhibition (EVS34) in 2021, the Natural Science Award of Chinese Association of Automation in 2018, and the Technical Invention Award of the Ministry of Education, China in 2019, respectively. His current research interests are data analysis of renewable energy systems, electric and fuel cell vehicles, and data science management of lithium-ion batteries.

Dr. Xin Lai received the Ph.D. degree in mechatronic engineering from the School of Mechanical and Energy Engineering, Tongji University, Shanghai, China, in 2013. He is currently an Associate Professor with the Department of Vehicle Engineering, School of Mechanical Engineering, University of Shanghai for Science and Technology, Shanghai, China. In recent years, he presided over two projects from the National Natural Science Foundation of China, and he has conducted extensive research and authored more than 50 journal papers (eight were selected as ESI highly cited papers) related to battery management. His research interests include battery management system particular for data science-based battery fault diagnosis and battery reutilization.

Abbreviations

ARD	Automatic relevance determination
BES	Battery-based energy storage
BPNN	Back-propagation neural network
CART	Classification and regression tree
CCCV	Constant current constant voltage
CEV	Commercial electric vehicle
CG	Comma gap
CM	Confusion matrix
CMC	Carboxymethyl cellulose
CNF	Carbon nanofiber
CRISP	Cross-industry standard process
DM	Data mining
DRT	Distribution of relaxation times
DT	Decision tree
DVA	Differential voltage analysis
ECM	Equivalent circuit model
EIS	Electrochemistry impedance spectroscopy
EKF	Extended Kalman filter
EM	Electrochemical model
EMD	Empirical mode decomposition
EoL	End-of-life
EV	Electric vehicle
FI	Feature importance
FN	False negative
FNN	Feedforward-neural network
FP	False positive
GMM	Gaussian mixture model
GPR	Gaussian process regression
HEV	Hybrid electric vehicle
HNBR	Hydrogenated nitrile butadiene rubber
ICA	Incremental capacity analysis

IMF	Intrinsic mode function
IPP	Intermediate process parameter
IPV	Intermediate product variable
ISC	Internal short circuit
KF	Kalman filter
LAM	Loss of active material
LCO	Lithium cobalt oxide
LFP	Lithium iron phosphate
LLI	Loss of lithium inventory
LMO	Lithium manganese oxide
LSTM	Long short-term memory
MAE	Mean absolute error
MC	Mass content
MCC	Multi-stage constant-current
ML	Machine learning
NCA	Lithium nickel cobalt aluminium oxide
NMC	Lithium nickel manganese cobalt oxide
NMP	N-methyl-2-pyrrolidone
NN	Neural network
NSGA-II	Non-dominated sorting genetic algorithm II
OCV	Open-circuit voltage
P2D	Pseudo-two-dimensional
PDE	Partial differential equation
PDF	Probability density function
PF	Particle filter
PHEV	Plug-in hybrid electric vehicle
PMOA	Predictive measure of association
PVAP	Predicted versus actual plot
PVDF	Polyvinylidene difluoride
RF	Random forest
RLS	Recursive least square
RMSE	Root mean square error
RNN	Recurrent neural network
RUBoost	Random undersampling boosting
RUL	Remaining useful lifetime
SBR	Styrene butadiene rubber
SEI	Solid electrolyte interface
SoC	State of charge
SoH	State of health
SoP	State of power
SoT	State of temperature
STLR	Solid-to-liquid ratio

SVM	Support vector machine
SVR	Support vector regression
TN	True negative
TP	True positive

Chapter 1
Introduction to Battery Full-Lifespan Management

As one of the most promising alternatives to effectively bypass fossil fuels and promote net-zero carbon emission target around the world, rechargeable lithium-ion (Li-ion) batteries have become a mainstream energy storage technology in numerous important applications such as electric vehicles, renewable energy storage, and smart grid. However, Li-ion batteries present inevitable ageing and performance degradation with time. To ensure efficiency, safety and avoid potential failures for Li-ion batteries, reliable battery management during its full-lifespan is of significant importance. This chapter first introduces the background and motivation of Li-ion battery, followed by the description of Li-ion battery fundamentals and the demands of battery management. After that, the basic information and benefits of using data science technologies to achieve effective battery full-lifespan management are presented.

1.1 Background and Motivation

1.1.1 Energy Storage Market

According to the statistics from the CNESA Global Energy Storage Projects Database, the global operating energy storage project capacity has reached 191.1GW at the end of 2020, a year-on-year increase of 3.4% [1]. As illustrated in Fig. 1.1, pumped storage contributes to the largest portion of global capacity with 172.5GW, a year-on-year increase of 0.9%. Electrochemical energy storage becomes the second-largest portion with a total capacity of 14.1GW. Among different electrochemical energy storage solutions, Li-ion batteries reach the capacity of 13.1GW, exceeding 10GW for the first time.

© The Author(s) 2022
K. Liu et al., *Data Science-Based Full-Lifespan Management of Lithium-Ion Battery*, Green Energy and Technology,
https://doi.org/10.1007/978-3-031-01340-9_1

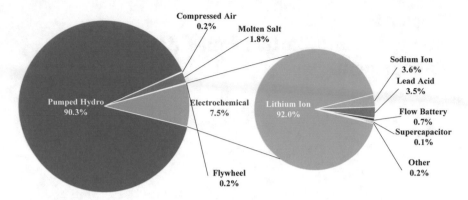

Fig. 1.1   2020 global energy storage market classification share, reprinted from [1], open access

China, the USA, Europe, and Australia are the leaders of the energy storage market. The new operating capacity of these countries accounted for over 86%, which has exceeded the GW-level new operating capacity.

China: Driven by the Chinese policies to encourage and require storage allocation in the energy, the largest installed capacity of new energy power generation in China exceeds 580 MW, a rapid increase of 438%. Furthermore, the establishment of "Carbon Peak" and "Carbon Neutral" targets in China also significantly boosts the leapfrog development of renewable energy and related battery-based energy storage.

In April 2021, the National Development & Reform Commission and the National Energy Administration issued the "Guiding Opinions on Accelerating the Development of New Energy Storage (Draft for Comment)" [2]. For the first time, the document clarifies the development goals of the energy storage industry. By 2025, the installed capacity of new energy storage capacity will reach more than 30 million kilowatts (30 GWh). As of 2020, the cumulative installed capacity of new electric energy storage in China has reached 3.28Gwh [3], a year-on-year increase of 91.2%, which also means that by 2025, the scale of the Chinese new energy storage market will be about 10 times larger than the level at the end of 2020.

USA: A breakthrough has been made in the deployment before the schedule in 2020, and the newly added operating capacity in the USA has doubled in comparison with that in 2019. The newly installed capacity is mainly concentrated in California, while LS Power and Vistra Energy added 250 MW/250MWh and 300 MW/1200MWh projects, respectively. The latter is the largest battery-based energy storage project in the USA and even the world. Besides, the deployment of large-scale 100 MW battery-based energy storage projects in Texas, New York, Florida, and other states has been accelerated.

The National Renewable Energy Lab uses the Regional Energy Deployment System (ReEDS) capacity expansion model to understand the complex dynamics involved with the future market potential for utility-scale energy storage in the contiguous USA [4]. The battery storage mandates enacted in Oregon, California,

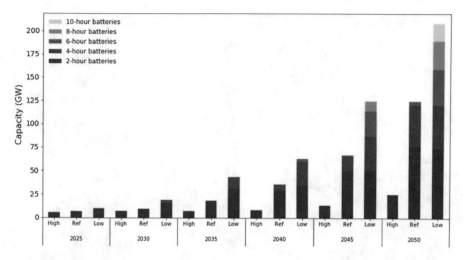

Fig. 1.2 Cumulative model-deployed battery storage with High, Reference, and Low battery capital costs. All scenarios show here use the dynamic assessment of storage capacity credit, reprinted from [4], with permission from Elsevier

New Jersey, New York, and Massachusetts are included. In total, these mandates require the model to build 1775 MW of batteries by 2020, 4685 MW by 2025, and 6555 MW by 2030. As illustrated in Fig. 1.2, the results of the "Reference" and "Low" battery cost scenarios generated by the ReEDS show a significant new battery storage deployment, with the deployment levels of 125 GW and 208 GW in 2050, respectively.

Europe: The implementation of the "Clean Energy for All" program has sent a significantly positive signal for the European energy storage market. This is reflected in the strong performance of the front-end energy storage market for electricity meters in the UK and the strong performance of the home energy storage market in Germany. The UK has cancelled project capacity restrictions, allowed more than 50 and 350 MW projects in England and Wales, and officially launched the construction of large-scale energy storage projects in the UK. With more than 300,000 household battery systems installed in Germany, COVID-19 has further stimulated consumer demand for energy flexibility, safety, and independence.

The European Union (EU) energy and climate policy aims to significantly cut CO_2 emissions in the power sector by 2030 and to establish a nearly carbon-free electricity sector by 2050. The role of transmission and energy storage in European decarburization towards 2050 provides support to the hypothesis that the EU energy and climate targets for 2050 will increase the capacity of intermittent power, storage technologies and international transmission lines, as illustrated in Fig. 1.3. In 2050, the investment in electric battery capacity will range from 80 to 351 GWh [5].

An assessment of European electricity arbitrage using storage systems shows that, in the near future, the most attractive European countries for the electricity

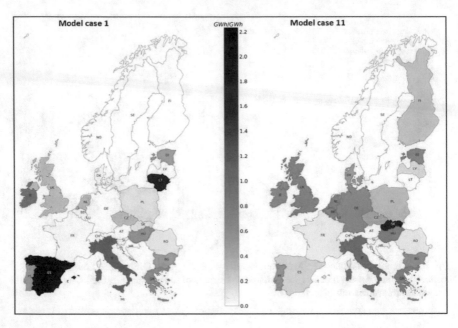

Fig. 1.3 Battery energy capacity per average hourly electricity demand in 2050 for the European countries, reprinted from [5], open access

arbitrage business are the UK and Ireland, with current Net Present Value close to −400,000 V, while Spain and Portugal might show the worst performances, their current Net Present Value are close to −800,000 V.

United Kingdom: The UK was the country with the largest new operating energy storage capacity in the European market, accounting for 44.6% of the total European continent in 2019. In 2019, the UK government signed a legally binding commitment to bring all greenhouse gas emissions to net-zero by 2050. Batteries will play a significant part in this transition, both in transport and renewable energy storage. Improving understanding of how batteries will age and how to design more efficient battery management system will aid the net-zero transition and reduce waste. Batteries and electric vehicles also represent part of the government's industrial strategy for the future of mobility and the mission to "Put UK at the forefront of the design and manufacturing of zero-emission vehicles" [6]. The UK automotive industry adds £18.6 billion to the UK economy, employees 823,000 people, and accounts for 14.4% of all UK goods exports [7]. To facilitate the transition to electrification, £1 billion is being invested in the Advanced Propulsion Centre (APC) and £246 million in the Faraday battery challenge.

Australian: The Australian Energy Market Operator (AEMO) reports that there are 85 big batteries with a total capacity of 18,660 MW in the planning pipeline [8]. How many of these projects can be realized will be a function of battery cost development, as well as the development of different revenue streams that batteries are enabled to provide. In Australia, batteries can provide revenue-generating services in various markets, most of which are ancillary and wholesale markets [9]. Virtual transmission lines, or avoided transmission investment, are emerging as a potential income stream for batteries.

1.1.2 Li-Ion Battery Role

Demand for Li-ion batteries to power electric vehicles and energy storage has seen exponential growth, increasing from just 0.5 gigawatt-hours in 2010 to around 526 gigawatt hours a decade later. Demand is projected to increase 17-fold by 2030.

According to the commissioned manufacturing capacity of Li-ion battery by plant location in Fig. 1.4, Asia dominates the Li-ion battery supply chain, especially China [10], where Chinese Li-ion battery manufacturer CATL is one of the world leaders in battery manufacturing in 2020, as illustrated in Fig. 1.5. China's success results from its sizeable domestic battery demand, control of more than 70% of the world's graphite raw material refining, and massive cell and cell component manufacturing capacity. Korea and Japan rank number two and three in the Li-ion battery supply chain. While both countries are among the leaders in battery and cell component manufacturing (LG Energy Solution, Samsung SDI, SK Innovation, Panasonic), they do not have the same influence in raw materials refining and mining as China.

Figure 1.6 illustrates the Li-ion battery cell manufacturing capacity by country or region. Obviously, China is the largest battery manufacturing country with 567 GWh, which is nearly ten times larger than the second one-the United States. Europe owns

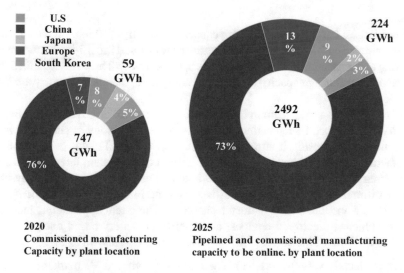

Fig. 1.4 Commissioned manufacturing capacity of Li-ion battery by plant location, 2020 and 2025, reprinted from [10], open access

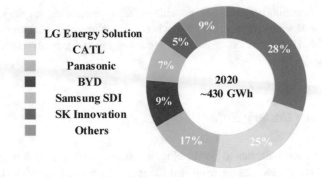

Fig. 1.5 2020 top battery manufacturers market shares in GWh

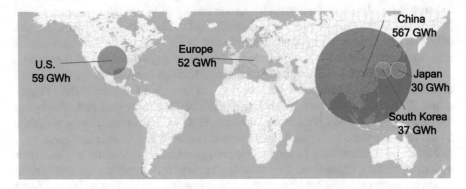

Fig. 1.6 Cell manufacturing capacity by country or region, reprinted from [10], open access

the third-largest battery manufacturing capacity. Apart from China, two Asian countries including South Korea and Japan present the fourth and fifth Li-ion battery manufacturing capacity with 37 GWh and 30 GWh, respectively. The detailed percentages of total manufacturing capacity for various battery components by country are listed in Table 1.1.

As analysed by Yole's team in the new Status of the Rechargeable Li-ion Battery Industry 2021 report, Li-ion battery has become the technology of choice for many applications. As a result, it attracts numerous players: R&D labs, cell component manufacturers, cell and battery pack manufacturers, and system integrators. Li-ion battery market is composed of multiple applications of battery technology, with slightly different targets and roles, further resulting in each application being best served by the specific Li-ion battery technology. Three real applications including electric vehicles, electronic devices, and stationary battery-based energy storage comprise the bulk of the current Li-ion battery market, as shown in Fig. 1.7.

Different electric vehicles (xEVs): The rapidly growing xEVs market consists of different types of EVs such as hybrid electric vehicle (HEV), plug-in hybrid electric vehicle (PHEV), full electric vehicle (EV), and commercial electric vehicle (CEV),

Table 1.1 Percentage of total manufacturing capacity of different countries for various components of Li-ion battery (data from [10], open access)

Country	Cathodes manufacturing (3 M tons) (%)	Anode manufacturing (1.2 M tons) (%)	Electrolyte solution manufacturing (339,000 tons) (%)	Separator manufacturing (1987 M m^2) (%)
United States	–	10	2	6
China	42	65	65	43
Japan	33	19	12	21
Korea	15	6	4	28
Rest of World	10	–	17	2

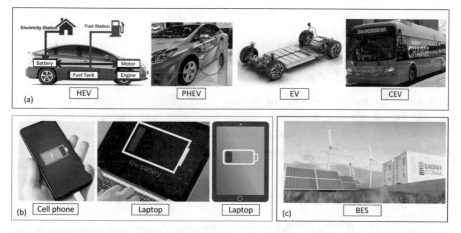

Fig. 1.7 Main applications of Li-ion battery: **a** different electric vehicles, **b** electronic device, **c** stationary battery-based energy storage

as illustrated in Fig. 1.7a, where Li-ion battery plays specific roles in different applications. For HEV, as it belongs to the traditional internal combustion engine-based vehicle, its propulsion system is combined with a small electric motor driven by batteries and these batteries are usually charged through regenerative braking. In this context, the capacity of the Li-ion battery is relatively smaller, further making its energy density and capital cost become less relevant. However, due to frequent braking of HEV, battery here requires to be charged and discharged powerfully. Therefore, Li-ion battery needs to have high power density, quick charging speed and long lifetime over thousands of cycles in HEV. In comparison with HEV, batteries within PHEV could be also charged through plugging into an external electricity source. The battery here generally presents much larger capacity to enable PHEV to drive fully electric for a short distance. In this context, Li-ion battery requires better energy density and lower capital cost, while its power density as well as lifetime become of less concern. For a full EV without any internal combustion engine, in order to deliver enough ranges for drivers, Li-ion battery generally needs low capital cost and high energy density. Besides, as EV could not fall back on the internal combustion engine anymore, Li-ion battery also needs to present high reliability and long service life over 1000 cycles. For CEV such as e-bus that battery systems are relatively larger and the effects of battery fault such as thermal runaway would become more severe, Li-ion battery here has increased safety requirements. Besides, as e-buses generally need to be charged frequently, the service life of Li-ion battery here also becomes more important than other xEV cases.

Electronic device: Li-ion battery is also widely used to support power/energy for electronic devices such as cell phones, laptops, and tablets, as shown in Fig. 1.7b. Li-ion battery presents the similar roles in all these electronic device applications to provide as much energy as possible in a compact form, so the volumetric energy density becomes the most crucial element. Besides, the cost of Li-ion battery in electronic devices is relatively smaller and users are generally willing to pay for high-performance Li-ion battery, cost here thus becomes the secondary important element. Furthermore, as electronic device application usually presents low drain, the power density of Li-ion battery here becomes less of concern.

Stationary battery-based energy storage (BES): BES is becoming a vital part to smooth the supply and demand of power generated from renewable energy such as wind sources and solar sources, as illustrated in Fig. 1.7c. In real applications, BES ensures the electricity transferred from renewable energy could be stored for further reutilization. Besides, it is able to also ensure that the peak in consumption is absorbed while backup could be provided without having to temporarily rely on fossil fuel power plants, further bringing positive environmental and economic impacts. There are different types of operating models for Li-ion batteries in BES applications. Based upon the requirements of Li-ion battery, these operating models could be divided across two axes as the frequency of discharge and the length of discharge, where the applications and key needs of Li-ion battery in four related quadrants of BES are illustrated in Fig. 1.8.

1.2 Li-Ion Battery and Its Management

1.2.1 Li-Ion Battery

Li-ion battery belongs to an electrochemical energy storage system, which generates a potential difference and allows current to flow through the circuit until the energy is exhausted. The first Li-ion battery was commercially introduced by Sony company in 1991. As illustrated in Fig. 1.9, Li-ion battery consists of three components including anode (negative electrode), cathode (positive electrode), and electrolyte. The active material is bonded to the metal fluid at both ends of the cell and electrically isolated

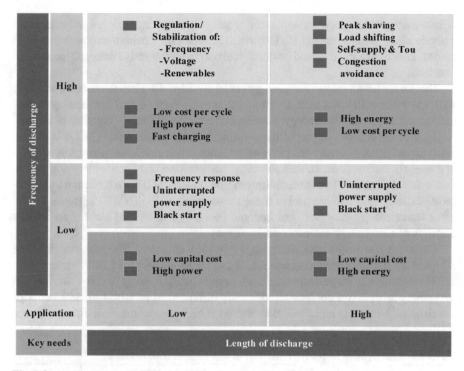

Fig. 1.8 Applications and key needs of Li-ion battery in four related quadrants of stationary battery-based energy storage

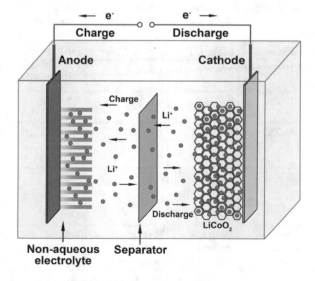

Fig. 1.9 Operating scheme of Li-ion battery

with a microporous polymer separator or gel polymer. Liquid or gel polymer electrolytes allow lithium ions (Li^+) to diffuse between the positive and negative electrodes. Li-ions are intercalated or deintercalated from the active material through an intercalation process.

The anode here mainly contains graphite. Besides, Li-Titanate anode combined with any other cathode is also developed to provide better safety and battery performance at the sacrifice of energy density. For cathode, it mainly consists of a metal oxide. Among different types of the cathode, lithium cobalt oxide (LCO) is able to offer higher energy density but presents a higher safety risk level, especially when it is damaged. In this context, this chemical composition has been widely adopted in consumer electronics. In contrast, lithium iron phosphate (LFP), lithium manganese oxide (LMO), and lithium nickel manganese cobalt oxide (NMC) batteries would offer lower energy densities, but become inherently safer. The electrolyte mainly consists of a lithium salt in an organic solvent.

Table 1.2 illustrates and summarizes various chemical compositions that are adopted for battery cathode electrode. It can be noted that for Li-ion battery with different materials, its performance such as the voltage, energy density, service life, and safety level could become significantly different. Most metal oxide electrodes are thermally unstable and could decompose at high-temperature conditions, further

Table 1.2 Rechargeable Li-ion batteries with various cathode compositions

Name	LCO	LMO	NMC	NCA	LFP
Full name	Lithium cobalt oxide	Lithium manganese oxide	Lithium nickel manganese cobalt oxide	Lithium nickel cobalt aluminium oxide	Lithium iron phosphate
Cathode	$LiCoO_2$	$LiMn_2O_4$	$LiNiMnCoO_2$	$LiNiCoAlO_2$	$LiFePO_4$
Cell voltage [V]	3.7	3.6	3.8	3.6	3.3
Cut-off voltage [V]	4.2	4.2	4.1–4.3	4.3–4.5	3.6
Energy density [Wh/kg]	150–250	100–170	150–220	200–250	80–140
Thermal runaway temperature [°C]	150	250	200	175	250
Common applications	Cellphone, camera, laptop	Hand tool, medical equipment	Hand tool, medical equipment	Vehicle	Hand tool, medical equipment
Comments	High energy density	Low internal impedance	High energy density Withstands rapid charge	High energy density High cell voltage	Long charge lifetime Withstands rapid charge

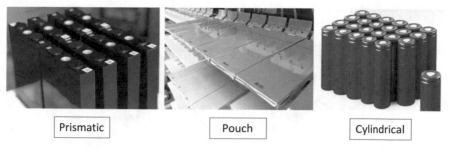

Fig. 1.10 Li-ion cell designs with various shapes

releasing oxygen to result in thermal runaway conditions. Among all these electrode chemical compositions, lithium manganese oxide (LMO) and lithium nickel manganese cobalt oxide (NMC) become the best candidates to compromise between performance and safety levels currently available on the Li-ion battery market.

Besides, Li-ion battery could be designed with different shapes including prismatic, pouch, and cylindrical, as shown in Fig. 1.10. Among these three designs, the prismatic battery cell design becomes the safest one because it is equipped with some mechanisms, such as safety functional layers, multi-layer partitions, safety vents, safety fuses, and overcharge safety devices.

1.2.2 Demands for Battery Management

(1) Battery management system market

Due to the dramatically increased requirements of battery being used in numerous applications such as transportation electrifications and smart grid energy storage, the global market of battery management system also grows rapidly with a compound annual growth rate of over 10%. Here the transportation sector is leading the main market growth of battery management system as a great number of EVs being manufactured and sold annually. For example, the global EV fleet stock was around 10.2 million in 2020, an increase of over 43% in comparison with that in 2019. Battery management plays a pivotal role in determining battery efficiency, performance and safety, especially for EV applications. Therefore, battery management system must be well equipped in an EV.

Figure 1.11 illustrates the growth rate of battery management system market around the world from 2021 to 2026 estimated by the Mordor Intelligence [11]. It can be seen that Asia–Pacific owns the biggest market share for battery management system, mainly due to the dramatically rising sale of EVs in countries like China and Japan. This dramatic increase of EVs is mainly caused by the extensive efforts of the governments to decrease greenhouse gas emissions. For example, China has become the biggest EV market around the world. The market share of Chinese EVs has risen from about 23% in 2015 to 44% in 2020, and by the end of 2020, about 4.5 million

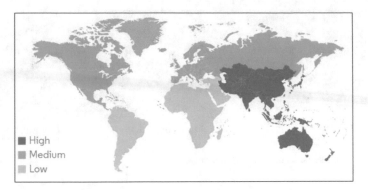

Fig. 1.11 Estimated growth rate of battery management system market around world from 2021 to 2026, reprinted from [11], open access

EVs have been deployed. Besides, the growth in demand for consumer electronic products would further increase the demand for battery management, because battery management system is increasingly integrated into consumer electronic products for security purposes. In this context, effective and reliable battery management systems or solutions are urgently required to meet the requirements of these battery-based electronic products.

(2) **Battery management system basic functions**

In Li-ion batteries, the key to longevity, efficiency, reliability and safety lies in the efficient management of battery under various operating levels. For transportation electrification applications such as EV, the main basic functions of battery management include: battery data acquisition, battery modelling, battery states estimation, battery ageing prognostics, battery fault diagnosis, battery charging, etc. [12, 13]. In general, all battery management solutions first rely on the quality of collected battery data. The sampling speed, measurement accuracy, and data pre-filtering are initial key elements to determine battery management performance. In this context, battery management system typically requires various types of sensors to measure the data of battery current, voltage and temperature. Furthermore, several battery internal state information such as the state of charge (SoC), state of power (SoP), and state of health (SoH) are difficult to be measured directly; therefore, various filtering and estimation algorithms such as Kalman filter and its variants, particle filter (PF), and neural network (NN) are employed to obtain information of these states [14, 15]. Besides, battery ageing information such as further capacity degradation trajectory and remaining useful life need to be predicted for reducing EV users' mileage anxiety [16]. Another high priority of battery management is fault diagnosis to ensure battery safety, which means that any critical failures must be detected or battery system must be shut down if a fault occurs [17]. To achieve fast, safe, and efficient charging management, battery charging strategies with the ability to handle various conflicting objectives and satisfy battery operating constraints also need to be carefully designed [18]. The current battery management system mainly monitors and controls batteries with fixed structures, which cannot provide full play to the optimal performance of battery systems. Han et al. designed a reconfigurable battery management system to allow the dynamic battery reconfiguration [19]. Dai et al. proposed a three layers-based battery management framework, including the foundation layer, algorithm layer and application layer [20]. These advanced battery management solutions are able to significantly improve battery safety, performance, and efficiency under various transportation electrification applications.

(3) **Battery management system challenges**

Nowadays, due to transportation electrification being the broadest application scenario for Li-ion battery, battery management solutions are mainly designed for EV applications where the battery capacity ages from 100 to 80%. To ensure effective battery performance under complex, volatile, and extreme operating cases, various battery operation management strategies have been designed to protect electrical vehicle battery against faulty operations and to optimize its charging or discharging dynamics. However, in comparison with battery operation management area with fruitful solutions, fewer works have been done so far on applying data science-based strategies to benefit battery manufacturing and reutilization.

For battery manufacturing, as battery initial performance would be directly determined by each intermediate stage within the manufacturing line, an effective battery management solution that can analyse the effects of manufacturing parameters on battery properties and optimize the manufacturing line is crucial. Besides, the battery could make up to 30% weight and cost of an EV [21], while contributing to more than 40% CO_2 emissions during the production of EV [22]. In light of this, efficient management of battery manufacturing towards high-quality battery and economic targets such as high manufacturing yield, low manufacturing cost, and less pollution is crucial and plays a pivotal role in the acceptance of battery. Currently, as battery manufacturing generally contains a number of chemical, mechanical, and electrical operations, and also generates numerous strongly coupled manufacturing parameters in the order of tens or hundreds, engineers often rely on the experiment experience, expert advice, trial and error solutions to analyse and manage their battery manufacturing line. These solutions would result in huge laborious and time consumptions, slow battery product development, inaccurate quality control, and difficulty in generating sustainable business cases for the technological introduction. Therefore, it is imperative to introduce advanced and smart solutions to manage battery manufacturing, and explore the correlation, interaction, interdependency of all relevant parameters, to improve battery manufacturing performance.

For battery utilization, on the one hand, a Li-ion battery is usually determined to be unsuitable for EV applications when its real capacity becomes less than 80% of its nominal value [23]. As a result, a large number of automotive batteries will be retired in the coming years [24]. For example, 250,000 metric tons of automotive batteries are predicted to hit their end-of-life (EoL) by 2025 [25]. The second-life battery has the potential to generate more than 200 GWh by 2030, with a global value of more than $30 billion, according to another report [26]. Again, even under the most optimistic estimates, 3.4 million kg of automotive batteries cells might end up in the waste stream by 2040 [27]. These numerous retired batteries containing volatile chemical elements would be released into the atmosphere if reutilization is not performed which will undergo both environmental and economic harm. On the other hand, in response to global climate change, many renewable and sustainable energy sources such as solar and wind have been adopted. However, due to the intermittent and time-varying existence of renewable energy sources, power fluctuates would be generated. This would significantly affect the grid performance, voltage stability, and reliability, making them become difficult to be processed into the grid. Based upon suitable battery reutilization solutions, this can be effectively mitigated if the generated energy from a renewable source is first deposited in a battery, and then converted by an appropriate power electronic converter topology to achieve the necessary grid voltage and frequency. In this context, giving such retired batteries a suitable reutilization solution, which is the management of batteries after they have reached 80% capacity would not only support the economy but also help to minimize total battery demand, resulting in a substantial reduction in the use of extracted chemical materials and significantly benefit many battery second-life applications such as grid energy storage [28].

Based upon the above discussions, battery full-lifespan from manufacturing, operation, and reutilization as a whole need to be carefully managed. With the rapid development of artificial intelligence and machine learning technology, data science-based tools stand out as the promising solutions for battery full-lifespan management, hopefully enabling us to overcome the major challenges dealing with different types of data from battery manufacturing, operation and reutilization. On the basis of this, a brand-new hologram to make full use of battery during full-lifespan could be formulated, further boosting the advancement of low-carbon technologies.

1.3   Data Science Technologies

To move data science-based tools applied to battery full-lifespan management efficiently, the systematic understanding and exploration of data science technology are required. In this context, data science-based tools must be properly explained and discussed in a way suitable for a broad audience.

1.3.1   What is Data Science

Data science is a practice of mining raw datasets with both structured and unstructured forms to identify specific patterns and extract meaningful insights from these data. It belongs to an interdisciplinary field, which mainly involves statistics, automation and engineering, computer science, machine learning, and new data-based technology to obtain insights from real data.

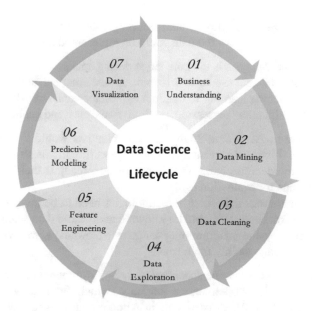

Fig. 1.12 Typical data science lifecycle

Figure 1.12 illustrates the typical lifecycle of data science, which includes seven main parts: business understanding, data mining, data cleaning, data exploration, feature engineering, predictive modelling, and data visualization. Business understanding mainly refers to the definition of relevant questions and objectives from the applications that require to be explored. For data mining, the necessary data needs to be gathered and scraped. Data cleaning involves the solutions to fix the inconsistencies within data and handle their missing values. For data exploration, data would be analysed visually to form the hypotheses of defined data science problem. For feature engineering, the importance and correlations of feature variables from data would be quantified and analysed. For predictive modelling, data science tools such as the machine learning models would be trained, validated, and adopted to make new predictions. For data visualization, conclusions will be reported to key stakeholders through various plots and interactive visualization tools.

In order to define data science task and clearly manage data science-based project, four main stages need to be carefully considered as:

Data architecture: The first stage in the data science pipeline workflow is to define data architecture. This requires data scientists to think through in advance how data users could make full use of data. Then data scientists also need to think about how to organize data to support different analyses and visualizations.

Data acquisition: The next stage is data acquisition which focuses on how to collect data from different sources such as experiments or real applications. Besides, various representing, transforming, and grouping solutions are all needed to help data scientists understand how the data could be represented before analysis.

Data analysis: Data analysis is the core stage during data science workflow. In this stage, data scientists would use various technical, mathematical, and statistical tools such as AI and machine learning to conduct exploratory and confirmatory analysis works such as classification, regression, predictive analyses, and qualitative analyses.

Insight conclusion: After data analysis, data scientists would communicate the obtained insights through data visualization and reporting in this stage. These could benefit the stakeholders to obtain useful conclusions, readjust their strategies, and generate new plans for evaluation again.

In addition, numerous data science technologies need to be involved in the data science pipeline workflow. It should be known that all these data science technologies are designed by the programming language. Nowadays, the widely utilized programming language mainly includes:

(1) **Python**: Based upon some specifical and easy-to-implement libraries, Python becomes a popular open-source language, which has been widely used by academia and industries particular in AI community. After being created in 1989, Python became a feasible programming language to offer numerous tools for manipulating datasets and analyse data science results easily and conveniently [134]. Besides, there exist some Python libraries that specifically focus on machine learning algorithm development, including and not limited to Scikit-Learn, Keras, and TensorFlow. Due to numerous programming language forums and websites having published many topics on the implementation of Python, popularity becomes another merit of Python [131, 132]. In general, Python-based data science solutions need to be executed under a cross-platform named integrated development environment (IDE) including Pycharm, Spyder, and Jupyter Notebook, where the friendly interface and the possibility of interacting Python with other programming languages become key elements that need to be considered.

(2) **MATLAB**: As an efficient programming language for technical computing, MATLAB is able to integrate computation, visualization, and programming in an easy-to-implement condition where data science issues and approaches could be expressed in the familiar mathematical notations. The basic data element of MATLAB is just an array without the need for dimensioning, further benefitting the computational effort of numerous data science computing issues particular for those with the formulation of matrix or vector. After being developed over a period of years, MATLAB has become a standard instructional tool for several introductory and advanced courses in automation engineering and data science particular in an academic environment. Besides, MATLAB features a family of toolboxes for users to explore and apply different technologies for their specific applications. For data science, many toolboxes such as the neural network toolbox, deep learning toolbox, optimization toolbox,

and statistics and machine learning toolbox have been widely used to solve particular classes of issues.

(3) **R language**: After being developed around the last decade of the twentieth century, R language also becomes particularly popular in the field of statistics science. In comparison with Python and MATLAB, R language is less applied to develop data science solutions. However, it is able to provide fully dedicated statistical libraries including the MASS, stats, fdata, and glmnet. Besides, users could conveniently search the details regarding how to apply each library and package based on the Comprehensive R Archive Network (CRAN).

(4) **C++ and Fortran**: C++ and Fortran belong to the modern pioneers of programming languages, which have been widely adopted as high-performance language. For the data science applications, several C++ libraries including the SHARK and MLPACK can be used to design machine learning. In comparison with Python, MATLAB, and R language, the implementation of data science code through C++ and Fortran would become more difficult as the necessary memory management is usually required.

1.3.2 Type of Data Science Technologies

Data science technologies mainly contain the types of supervised, unsupervised, and semi-supervised approaches. For the supervised ones, certain variables need to be defined as input and output terms before a related dataset is employed. On the contrary, the definition of input and output terms is missing for the unsupervised data science methods, whose target is to automatically discover patterns in the dataset. Semi-supervised methods are somewhere in between supervised and unsupervised ones, which would apply datasets containing both labelled and unlabelled data.

Supervised data science approaches could be further divided into two main categories including regression methods and classification methods. The data science regression model analyses and outputs data in terms of continuous values, while the data science classification model will analyse and output classes in terms of discrete values. These classes utilized for supervised data science approaches can come from the operator or from the unsupervised data science methods. For battery management applications, apart from the type adopted, classical data science approaches heavily depend on the data and are quite unrelated to physics, which means that they can be aimed at, for example, determining the underlying mapping among various variables, rather than providing any physical explanation of such mapping. However, through coupling battery physical elements, physical-driven data science methods also exist.

Figure 1.13 shows several most utilized data science methods in battery research and development. It should be known that all these methods have been adopted in the applications of battery full-lifespan management.

Neural network (NN): As illustrated in Fig. 1.13a, NN is proposed to mimic human brain activities through using the processing unit of artificial neurons arranged in the input layer, output layer as well as hidden layers. After a pre-processing stage,

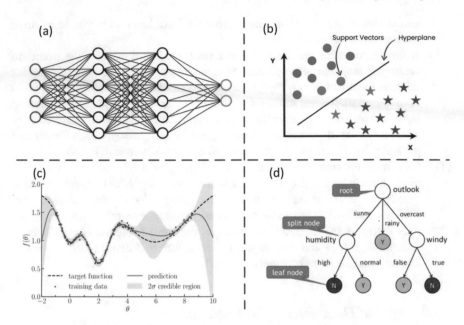

Fig. 1.13 Several typically utilized data science methods in battery research and development: **a** neural network, **b** support vector machine, **c** Gaussian process regression, **d** tree-based solution

data will be inputted into the input layer with a predefined input matrix. Then the neurons in hidden layers contain the mathematical function to generate output across neurons and could be expressed through using a weighted linear combination being wrapped in the activation functions [29]. In theory, The larger the neuron weight, the sensitivity to this specific input would become higher. Finally, the output layer would output the predicted values of NN. For the NN training process, the parameters here are mainly optimized by considering the amount of hidden layers, the number of neurons within each layer, the interconnected neuron weights, and the activation function types.

To date, two different types of NNs are widely utilized in battery management applications, including the feedforward-neural network (FNN) and recurrent-neural network (RNN). In the former, the data would travel in one direction only. After equipping feedback connection with FNN, RNN is derived. By involving the recurrent links, RNN is capable of keeping and updating the previous information for a period of time, making it a promising tool to capture the sequential correlations in battery management applications. For example, as battery ageing process usually contains hundreds of cycles, while the ageing information among these cycles is highly correlated. It is thus meaningful to extract and store these correlations for accurate battery lifetime prognostics. Besides, as it is able to capture the long-term dependency of data, RNN thus becomes an effective tool for capturing and updating sequential information during battery management.

One obvious benefit of NN is its capability of studying from experience and can adapt to varied situations. However, NN requires a large amount of battery application data to train and verify, and its accuracy would be heavily influenced by the training way and data quality. Furthermore, the computational effort is still a bottleneck for its large-scale application in battery management and NN structure also plays a pivotal role in determining its performance. In this context, NN optimization still remains an open technical problem. In general, the NN structure is determined by time-consuming trial and error. In light of this, some optimization approaches such as the two-stage stepwise identification method [30] could be adopted to optimize the NN structure for battery management applications.

Support vector machine (SVM): According to the kernel functions, SVM belongs to a supervised data science tool. It could perform both classification and regression tasks by searching the hyperplane separating classes with a maximal margin, as shown in Fig. 1.13b. SVM could adopt kernels to handle nonlinear problems by transforming the nonlinear issue with a low-dimensional space into a linear problem with a higher-dimensional space [31]. In theory, the prediction of SVM is based on several functions defined over the input space, and learning is a process to infer this function's parameters. SVM could make predictions with the function as:

$$y(x) = \sum_{n=1}^{N} \omega_n K(x, x_n) + \varepsilon \tag{1.1}$$

where ω_n represents the weights to connect feature space into output, $K(\cdot)$ stands for kernel function, and ε represents the independent noise.

SVM is particularly appealing for its ability to handle training datasets with small size. Here the number of support vectors would increase when the size of the training dataset becomes larger. To enhance the stability and robustness of SVM under large-scale training dataset size, decremental and incremental solutions [32] could be adopted by integrating relevant data samples for SVM training and ignoring irrelevant parts. However, the computational effort would be also increased in this process.

Gaussian process regression (GPR): Deriving from the Bayesian framework, GPR-based data science models have been widely adopted in battery prognostic applications due to their superiority in terms of being flexible, nonparametric, and probabilistic [33]. GPR is also a kernel-based data science approach, which is capable of realizing predictions combined with prior knowledge as well as providing variance around its mean prediction point to express the associated uncertainties, as shown in Fig. 1.13c. Here the Gaussian process can be regarded as the collection of a limited amount of random variables that present the joint multi-variate Gaussian distribution. In theory, the performance of GPR is significantly sensitive to the kernel functions, so the kernel functions need to be carefully designed for achieving high prediction accuracy. Battery application is usually complicated and would be affected by many impact elements. The single kernel function will easily lead to unreliable predictions for nonlinear mappings with multi-dimensional input terms.

In this context, an isotropic kernel with advanced structures such as the automatic relevance determination could be utilized. Furthermore, hyperparameters optimization of kernel functions within GPR is crucial as improper hyperparameters are easy to lead overfitting issue. To ameliorate this, minimizing the negative log marginal likelihood is generally adopted [34].

Tree-based solutions: Tree-based solutions are the decision-support data science by adopting the flowchart-like model to achieve classification or regression, as illustrated in Fig. 1.13d. Many tree-based solutions such as decision tree (DT), random forest (RF), and boosting-based approaches have been successfully utilized in the applications of battery management. The basic idea of DT is to divide a complicated prediction issue into many smaller ones based on a tree structure. In this way, each node within a DT could represent a small subissue, while DT as a whole could constitute a solution to the overall issue [35]. For DT training, data would be first injected in a root (i.e. the first node of DT). After that, the input term which could best discriminate between the output would be searched. That is, which value (V_i) of which input terms could split the initial dataset in such a way to separate as many outputs as possible would be searched to minimize related errors. This would result in the node being divided into two paths: one for values of the selected input items lower than V_i while another for values larger than V_i. The iteration of this process would result in a series of paths linking possible inputs to a certain output. One obvious benefit of DT is that an easy-to-understand representation of the links between input and output items could be generated. However, due to the too simple structure, DT is difficult to achieve high performance particularly for highly nonlinear applications. To handle this, many ensemble learning solutions such as RF are designed through combining DTs to improve overall prediction performance. The logic of RF can be summarized as that if the single DT cannot provide results with enough accuracy, the result through averaging all outputs from numerous DTs with a bagging solution would result in more accurate predictions [36]. This could bring significant improvement of prediction performance, further making RF become competent to solve highly nonlinear issues. Besides, boosting-based approaches also adopted some DTs to decrease both bias and variance of derived data science models, while the related prediction accuracy could be improved. The main difference here is the adopted sampling approach, where bagging and boosting solutions are differentiated by the procedure utilized for the training process.

1.3.3 Performance Indicators

After establishing data science-based solutions for different battery management applications, a key task is to adopt suitable performance indicators for evaluating the performance of these data science-based solutions. These performance indicators can be divided into two main categories to evaluate the regression and classification results, respectively.

(1) Data science regression model

To quantify and evaluate the accuracy of devised data science-based regression models in various battery management cases, the following three typical performance indicators are usually utilized.

Mean absolute error (MAE): Supposing N is the total number of regression samples, Y_i represents the actual reference value while $\hat{Y}_i$ stand for the output predicted from data science regression models, then the MAE could be obtained to evaluate the repression accuracy as:

$$\text{MAE} = \frac{1}{N} \sum_{i=1}^{N} \left| Y_i - \hat{Y}_i \right| \tag{1.2}$$

Root mean square error (RMSE): According to the same character definition, RMSE is another typical performance indicator to present the deviations between the predicted output and actual reference value as:

$$\text{RMSE} = \sqrt{\frac{1}{N} \sum_{i=1}^{N} \left(Y_i - \hat{Y}_i \right)^2} \tag{1.3}$$

R^2 **value:** Supposing $\overline{Y}_i$ is the mean value of all response outputs, R^2 value is also a typical performance indicator to reflect how closely the outputs from regression model could match well with the actual reference values as:

$$R^2 = 1 - \sum_{i=1}^{N} \left(Y_i - \hat{Y}_i \right)^2 \Big/ \sum_{i=1}^{N} \left(Y_i - \overline{Y}_i \right)^2 \tag{1.4}$$

For the regression applications, when the outputs predicted from models get close to the real experimental ones, MAE and RMSE present to be close to 0, while R^2 would get close to 1, indicating that a data science regression model is capable of explaining all the variability of target outputs.

(2) Data science classification model

For the classification cases, to quantify and evaluate the performance of the designed data science classification model, several performance indicators including the confusion matrix, macro-precision, macro-recall, and macro-F1-score are generally adopted.

Precision rate (P_{rate}): Supposing positive corresponds to the class of interest while negative corresponds to other classes, four basic measures including the true positive (TP), false positive (FP), true negative (TN), and false negative (FN) could be formulated for each class. Then for the class of interest C_i ($i = 1, \ldots, N_c$, N_c is the number of classes), its P_{rate} could be obtained to quantify the correct classification result of this class as:

$$P_{\text{rate}} = \text{TP}/(\text{TP} + \text{FP}) \tag{1.5}$$

Recall rate (R_{rate}): R_{rate} is able to quantify the rate of all fraud cases of this class as:

$$R_{\text{rate}} = \text{TP}/(\text{TP} + \text{FN}) \tag{1.6}$$

F-measure (F-measure): F-measure can reflect the harmonic mean of precision as well as recall of this class as:

$$F\text{-measure} = \frac{2 \times P_{\text{rate}} \times R_{\text{rate}}}{P_{\text{rate}} + R_{\text{rate}}} \tag{1.7}$$

Overall correct classification rate (OCC_{rate}): OCC_{rate} that reflects the proportion of correctly classified observations out of all the observations could be obtained by:

$$\text{OCC}_{\text{rate}} = \frac{\text{TP}_{\text{all}} + \text{TN}_{\text{all}}}{N} \tag{1.8}$$

where $\text{TP}_{\text{all}} + \text{TN}_{\text{all}}$ stands for all the correctly classified outputs from data science classification model, N represents the total amount of observations.

Confusion matrix (CM): According to the aforementioned metrics, a CM with $M+1$ rows and $M + 1$ columns can be formulated to reflect the performance of multiple class-based classification model. Here each row within CM is able to reflect the predicted output classes while each column stands for the actual target classes. The elements on the primary diagonal of CM reflect the correctly classified results, while other elements stand for the incorrectly classified conditions. The $M + 1$th column and $M + 1$ th row stand for the $P_{\text{rate}}(C_i)$ and $R_{\text{rate}}(C_i)$ of each class, respectively. The last element in the right-bottom corner is the OCC_{rate}.

macro$P$, macro$R$, and macro$F1$: Supposing each class has a $P_{\text{rate}}(C_i)$, $R_{\text{rate}}(C_i)$, and F-measure (C_i), then various overall performance indicators including the macro-precision (macro$P$), macro-recall (macro$R$), and macro-F1-score (macro$F1$) could be obtained to evaluate the overall classification performance of data science classification model as:

$$
\left\{
\begin{array}{l}
\mathrm{macro}\,P = \sum_{i=1}^{N_c} P_{\mathrm{rate}}(C_i)/N_c \\[2mm]
\mathrm{macro}\,R = \sum_{i=1}^{N_c} R_{\mathrm{rate}}(C_i)/N_c \\[2mm]
\mathrm{macro}\,F1 = \sum_{i=1}^{N_c} F\text{-measure}(C_i)/N_c
\end{array}
\right.
\tag{1.9}
$$

For the classification applications, when the classes outputted from a model match observations as much as possible, $\mathrm{macro}\,P$, $\mathrm{macro}\,R$, and $\mathrm{macro}\,F1$ would get close to 1, indicating that the data science classification model is able to perform high accurate classification.

Based upon the aforementioned classical performance indicators, the performance of data science-based battery management solutions can be quantified and evaluated.

1.4 Summary

This chapter first introduces the background and motivation of Li-ion battery. It outlines the role of Li-ion battery in the energy storage market of several leading countries. Three applications to comprise the bulk of the current Li-ion battery market including the electric vehicle, electronic device, and stationary battery-based energy storage are also introduced. Then, it describes the fundamental of Li-ion battery and the demands of battery management. Apart from battery operation management with fruitful solutions, the management of both battery manufacturing and reutilization is still in its infancy. In this context, with the rapid development of artificial intelligence and machine learning, data science-based solutions become a promising way to handle various key challenges of battery full-lifespan management. After that, this chapter reviews the basic information on data science lifecycle and widely utilized programming language and outlines the popular data science technologies used in battery full-lifespan management and corresponding performance indicators for result evaluation. It emphasizes the necessity and benefits of using data science technologies to manage batteries, while also guiding the design and development of data science-based tools for effective battery full-lifespan management.

References

1. China Energy Storage Alliance: Energy Storage Industry White Paper 2021. http://en.cnesa.org/
2. Guiding Opinions on Accelerating the Development of New Energy Storage (Draft for Comment). http://www.nea.gov.cn/139896047_16230559585841n.pdf
3. National Energy Administration: http://www.nea.gov.cn/2021-07/30/c_1310097494.htm
4. Frazier AW, Cole W, Denholm P, Greer D, Gagnon P (2020) Assessing the potential of battery storage as a peaking capacity resource in the United States. Appl Energy 275:115385

5. Golombek R, Lind A, Ringkjøb H-K, Seljom P (2022) The role of transmission and energy storage in European decarbonization towards 2050. Energy 239:122159
6. HM Government Industrial Strategy—The road to Zero, 2018
7. SMMT UK Automotive. www.smmt.co.uk/industry-topics/uk-automotive/
8. Martin N, Rice J (2021) Power outages, climate events and renewable energy: Reviewing energy storage policy and regulatory options for Australia. Renew Sust Energ Rev 137:110617
9. Large-scale battery storage. Australian Renewable Energy Agency. Knowledge Sharing Report. ARENA, 2019. https://arena.gov.au/assets/2019/11/large-scale-battery-storage-kno wledge-sharing-report.pdf
10. National Blueprint for Lithium Batteries 2021–2030. https://www.energy.gov
11. Mordor Intelligence,: Battery management system market- growth, trends, Covid-19 impact, and forecasts (2021–2026). https://www.mordorintelligence.com/industry-reports/battery-man agement-system-market
12. Lu L, Han X, Li J, Hua J, Ouyang M (2013) A review on the key issues for lithium-ion battery management in electric vehicles. J Power Sources 226:272–288
13. Liu K, Li K, Peng Q, Zhang C (2019) A brief review on key technologies in the battery management system of electric vehicles. Front Mech Eng 14(1):47–64
14. Hu X, Feng F, Liu K, Zhang L, Xie J, Liu B (2019) State estimation for advanced battery management: key challenges and future trends. Renew Sust Energ Rev 114, 109334
15. Wang Y, Tian J, Sun Z, Wang L, Xu R, Li M, Chen Z (2020) A comprehensive review of battery modeling and state estimation approaches for advanced battery management systems. Renew Sust Energ. Rev 131:110015
16. Hu X, Xu L, Lin X, Pecht M (2020) Battery lifetime prognostics. Joule 4(2):310–346
17. Hu X, Zhang K, Liu K, Lin X, Dey S, Onori S (2020) Advanced fault diagnosis for lithium-ion battery systems: a review of fault mechanisms, fault features, and diagnosis procedures. IEEE Ind Electron Mag 14(3):65–91
18. Tomaszewska A, Chu Z, Feng X, O'kane S, Liu X, Chen J, Ji C, Endler E, Li R, Liu L (2019) Lithium-ion battery fast charging: a review. ETransportation 1:100011
19. Han W, Wik T, Kersten A, Dong G, Zou C (2020) Next-generation battery management systems: dynamic reconfiguration. IEEE Ind Electron Mag 14(4):20–31
20. Dai H, Jiang B, Hu X, Lin X, Wei X, Pecht M (2021) Advanced battery management strategies for a sustainable energy future: multilayer design concepts and research trends. Renew Sust Energ Rev 138:110480
21. Nykvist B, Nilsson M (2015) Rapidly falling costs of battery packs for electric vehicles. Nat Clim Change 5(4):329–332
22. Duffner F, Kronemeyer N, Tübke J, Leker J, Winter M, Schmuch R (2021) Post-lithium-ion battery cell production and its compatibility with lithium-ion cell production infrastructure. Nat Energy 6(2):123–134
23. Winslow KM, Laux SJ, Townsend TG (2018) A review on the growing concern and potential management strategies of waste lithium-ion batteries. Resour Conserv Recycl 129:263–277
24. Lybbert M, Ghaemi Z, Balaji A, Warren R (2021) Integrating life cycle assessment and elec-trochemical modeling to study the effects of cell design and operating conditions on the environmental impacts of lithium-ion batteries. Renew Sust Energ Rev 144:111004
25. Stringer D, Ma J (2018) Where 3 million electric vehicle batteries will go when they retire. Bloom. Businessweek 27
26. Lovell J. Storage: retirement home for old EV batteries? https://www.energycouncil.com.au/analysis/storage-retirement-home-for-old-ev-batteries/
27. Casals LC, García BA, Canal C (2019) Second life batteries lifespan: rest of useful life and environmental analysis. J Environ Manage 232:354–363
28. Deng Y, Zhang Y, Luo F, Mu Y (2020) Operational planning of centralized charging stations utilizing second-life battery energy storage systems. IEEE Trans Sustain Energy 12(1):387–399
29. Abiodun OI, Jantan A, Omolara AE, Dada KV, Mohamed NA, Arshad H (2018) State-of-the-art in artificial neural network applications: a survey. Heliyon 4(11):e00938

30. Li K, Peng J-X, Bai E-W (2006) A two-stage algorithm for identification of nonlinear dynamic systems. Automatica 42(7):1189–1197
31. Suthaharan S (2016) Machine learning models and algorithms for big data classification. Springer, Berlin, pp 207–235
32. Chen Y, Xiong J, Xu W, Zuo J (2019) A novel online incremental and decremental learning algorithm based on variable support vector machine. Clust Comput 22(3):7435–7445
33. Schulz M, Speekenbrink A, Krause A (2018) A tutorial on Gaussian process regression: Modelling, exploring, and exploiting functions. J Math Psychol 85:1–16
34. Lucu M, Martinez-Laserna E, Gandiaga I, Camblong H (2018) A critical review on self-adaptive Li-ion battery ageing models. J Power Sources 401:85–101
35. Biau G, Scornet E (2016) A random forest guided tour. TEST 25(2):197–227
36. Altman N, Krzywinski M (2017) Ensemble methods: bagging and random forests. Nat Methods 14(10):933–935

Chapter 2
Key Stages for Battery Full-Lifespan Management

As a classical electrochemical component, Li-ion battery ages with time, losing its capacity to store charge and deliver it efficiently. In order to ensure battery safety and high performance, it is vital to design and imply a series of management targets during its full-lifespan. This chapter will first offer the concept and give a systematic framework for the full-lifespan of Li-ion battery, which can be mainly divided into three stages including the battery manufacturing, battery operation, and battery reutilization. Then key management tasks of each stage would be introduced in detail.

2.1 Full-Lifespan of Li-Ion Battery

Figure 2.1 gives a schematic diagram of battery full-lifespan, which consists of three main stages: battery manufacturing, battery operation, and battery reutilization. Here, battery manufacturing is related to the process that the battery is manufactured, which can be further divided into material preparation, electrode manufacturing, and cell manufacturing. As battery manufacturing could directly affect the properties of intermediate products such as the electrode volume ratio and thickness, further determining the initial performance of battery products, the battery manufacturing line as a whole needs to be well managed to produce suitable electrode architectures, assuring good electrical and ionic conductivities with the current collector despite low additive volume ratio. After manufacturing a battery with a capacity of 100%, it would be operated under different applications with high energy or power requirement, where the capacity value of battery will degrade gradually. As Li-ion batteries with a capacity value above 80% are suitable for supporting power and energy of the electric vehicle (EV), the process that battery capacity degrades from 100 to 80% is defined as the second stage of battery full-lifespan in this book, which is named as the operation stage. During battery operation management, advanced strategies for numerous aspects such as battery state estimation, battery lifetime prognostics, battery fault

K. Liu et al., *Data Science-Based Full-Lifespan Management of Lithium-Ion Battery*, Green Energy and Technology, https://doi.org/10.1007/978-3-031-01340-9_2

Fig. 2.1 Schematic diagram of battery full-lifespan

diagnosis, and battery charging need to be carefully designed for ensuring battery efficiency, performance, and safety. When the capacity of battery degrades below 80% of its nominal capacity, it becomes unsuitable for EV applications. In this case, battery needs to be retired from EVs and enters into the third stage of battery full-lifespan, which is called the reutilization stage in this book. It should be known that the batteries retired from EVs also have high economic or environmental values, which can be used in many battery second-life applications such as smart grid energy storage and low-speed transportation. In this context, the best way to dispose of the retired batteries is to perform echelon utilization first and then recycle materials, which can maximize the value of batteries and promote the healthy and sustainable development of batteries. Therefore, during battery reutilization management, the residual value of retired batteries will be evaluated by using the battery historical data or test data. Based upon the obtained residual value information, these batteries will be then sorted and regrouped for safe echelon utilization. The residual value such as the capacity of these batteries will continue decreasing during battery reutilization process. When the gradually reduced residual value of batteries cannot meet the requirements of battery different second-life applications, various material recycling solutions will be carried out to help recover the valuable materials, realize the recycling of valuable resources, reduce the impact of waste treatment on the environment, and reduce the development and consumption of natural resources. Then these recycled materials can be utilized for battery manufacturing again.

Based on the above discussion, the manufacturing, operation, and reutilization constitute a closed-loop circle for the full-lifespan of the battery. To ensure high performance of battery and make full use of battery during its full-lifespan, all these processes within battery lifespan should be well managed. On the other hand, during battery full-lifespan, lots of data related to battery behaviours would be generated. With the help of these valuable data and state-of-the-art data science tools, various data science-based solutions can be designed and developed to meet different management requirements of battery full-lifespan.

2.2 Li-Ion Battery Manufacturing

Manufacturing is the first stage of Li-ion battery lifespan. The main task in this stage is to produce affordable batteries with high performance and low manufacturing costs. As over 40% cost of an EV is spent on the battery manufacturing stage, a key issue that limits the wider applications of Li-ion battery lies in the improvement of its manufacturing management. Besides, the processes within battery manufacturing line play a vital and direct role in determining battery qualities, which should be well monitored and analysed. However, as battery manufacturing contains many material, chemical, mechanical, electrical operations and often generates strongly coupled variables in the order of tens or hundreds, the current mainstream solutions to analyse its feature variables are still based on trial and error, which is laborious and time-consuming. In this context, a reliable data science strategy to decouple correlations among various manufacturing variables and further explore impact variables towards resulting manufactured battery performance is challenging but urgently needed.

2.2.1 Battery Manufacturing Fundamental

Figure 2.2 illustrates a general framework of battery manufacturing. Since the 1990s, the manufacturing of Li-ion battery follows similar processes [1]. In general, after preparing materials, the process of battery manufacturing starts with mixing these materials to generate a homogenous slurry for coating, then the coating products would be dried, calendared, and cut into a specific size for assembling. After that, the assembled cell requires electrolyte filling, followed by the formation step to finish the manufacturing of a battery cell [2].

Materials preparation: In general, battery electrode materials contain the active material, conductive additives, and binder. For the cathode, common active materials include $LiNi_xCo_yMn_zO_2$ (NMC), $LiNi_xCo_yAl_zO_2$ (NCA), $LiMn_2O_4$, and $LiFePO_4$ (LFP). For the anode, graphite is usually used, sometimes with added silicon, and in some cells $Li_4Ti_5O_{12}$ (LTO). The typical conductive additives include carbon black, and carbon nanofibers. The polymer binder would provide the electrode mechanical integrity, which is usually the polyvinylidene difluoride (PVDF) for cathode and the mixture of carboxymethyl cellulose (CMC) and styrene-butadiene rubber (SBR) for

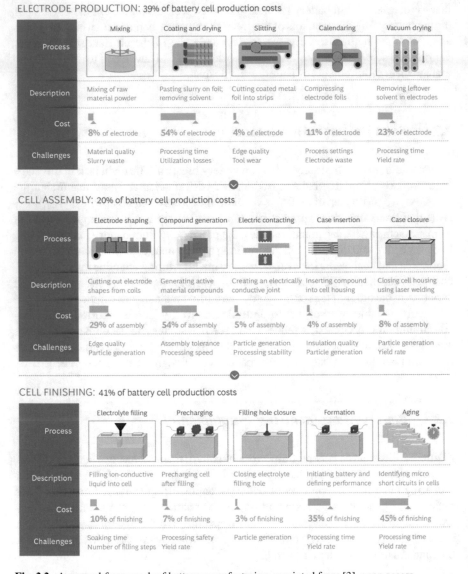

Fig. 2.2 A general framework of battery manufacturing, reprinted from [3], open access

anode [4]. The electrode materials are dispersed in the solvent, normally N-methyl-2-pyrrolidone (NMP) for PVDF cathodes, and water for graphite anodes, to generate the homogenous slurry.

Electrode manufacturing: Battery electrode manufacturing chain, as illustrated in Fig. 2.3c, could be further divided into several individual stages including mixing, coating, drying, calendaring, and cutting or slitting. For the mixing stage, the main

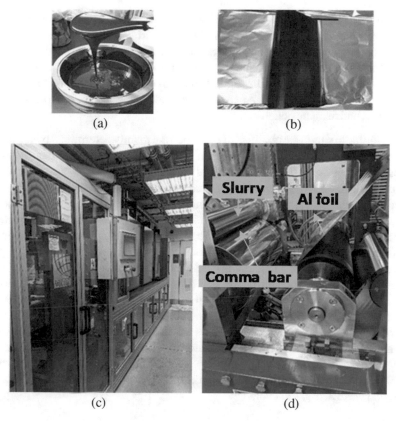

Fig. 2.3 Battery electrode manufacturing line: **a** slurry mixture, **b** coated foil, **c** electrode manufacturing machine, **d** coating machine

purpose here is to deagglomerate and disperse solid components into a liquid phase to generate homogenous slurry [5], as shown in Fig. 2.3a. For the coating stage with a specific machine as illustrated in Fig. 2.3d, the slurry would be coated onto the surface of current collectors (Cu and Al foil for anode and cathode coatings in Fig. 2.3b, respectively). It should be known that defects would occur in both mixing and coating stages, such as pinholes, agglomerates, and non-uniformities, which would highly affect electrode final electrical performance [6]. These defects could be caused by insufficient mixing, slurry degassing, and hardware malfunction in extreme conditions. Besides, the microstructure formed in the drying process would also determine both mechanical and electrochemical properties of electrode [7]. Typical drying generally consists of two stages. For the first stage, the solvent would be evaporated from the top surface of wet coating at a constant rate, leading it to shrink. For the second stage, solvent from pores would be emptied with an evaporation rate gradually reducing to zero [8]. Particles including active materials and conductive additives could be rearranged in the first stage. Excessive temperatures and drying rates would cause the binder to accumulate at the coating surface, further resulting in the electrode delamination [9]. For the calendaring stage, the coating product would be compacted by the mechanical pressure applied through two cylindrical rolls. Electrode thickness would be reduced to improve energy density, adhesion and form further networks within electrode. The final electrode microstructure could be reflected by its tortuosity [10]. After that, electrode would be cut into proper shapes for coin, cylindrical, or pouch cells.

Cell manufacturing: The qualities of electrode are mainly determined by the mixing to cutting stages and are difficult to be enhanced in the later battery manufacturing stages. In the case of cell manufacturing, electrodes and separators are stacked alternatively in a formed cell, then the packed cell will be filled with electrolyte and sealed. Both electrolyte filling and formation are critical to determining the performance of manufactured cell [11, 12]. The electrode porous network offers the internal surface area in contact with electrolyte. The interconnected micropores (smaller than 2 nm) and mesopores (2–50 nm wide) get filled with electrolyte and provide pathways for ionic diffusion [13]. Battery cell formation and testing are the final but one of the longest stages. During the cell formation stage, a stable solid electrolyte interface (SEI) would be formed through the consumption of active lithium and electrolyte material. This SEI layer could provide protection against further irreversible electrolyte consumption and damage to the active material particle [14]. Battery cells would be degassed to eliminate the generated products and then tested. Electrochemical tests are customized for different applications, but in principle, current/voltage would be measured to determine the manufactured cell performance such as capacity, energy/power density, life cycle.

2.2.2   Identifying Manufacturing Parameters and Variables

As illustrated in Fig. 2.4, the complexity of battery manufacturing management arises from each intermediate stage leaving its fingerprint on the intermediate products by the intermediate product variables (IPVs) influencing the subsequent steps, as well as the final properties of the manufactured battery. Each intermediate process has its own intermediate process parameters (IPPs) to determine IPVs' properties within this production stage. For the entire battery manufacturing line, as there exist numerous interdependent stages and each stage has its own IPPs, the total order of IPPs could reach the level of hundred. To explore IPPs and their effects on IPVs, some key IPPs are treated as control parameters or factors in the design of experiments and the corresponding IPVs as that step's responses.

Materials and mixing: Electrode material formulation (i.e. the amount of active material, additives, and binder), precursors, route, and mixing equipment are the key IPPs of the mixing stage. In general, electrode materials and composition are selected based on the performance, cost, safety [15, 16], and the chemical compatibility of compounds to ensure slurry stability. It should be known that active material plays a

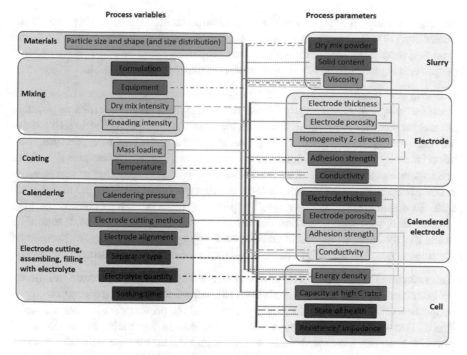

Fig. 2.4 Generalized battery manufacturing process showing intermediate process parameters (IPPs) and intermediate process variables (IPVs)

vital role in determining electrode gravimetric capacity. The power of active material will influence particle size and its distribution would determine slurry viscosity and compaction in coating and calendaring stages. Then the particle with too small size would lead to the high reactivity with electrolyte, but too large size will become more difficult to lithiated and delithiated bulk. In general, the mixed slurry would be characterized by viscosity (measured at a relevant value of shear rate with rotational rheometer), complex shear modulus (measured with oscillation rheometer), solid content, conductivity (when coated on the insulating substrate), surface, and interfacial tension, contact angle.

In regards to discovering an optimum formulation, the roles of active material (charge storage), the conductive additives (creating conductive paths for the Li-ions), and the binder (providing mechanical integrity) have to be balanced. A possible solution to optimize the mixing stage involves the testing of IPPs such as formulation, solid content, mixing equipment, dry mix intensity, power input, type of mixers, mixing speed, and kneading intensity. These IPPs are quantitative factors for which the experimenters can determine a relevant range with its corresponding number of levels. The corresponding IPVs caused by the IPPs in the materials preparation and mixing stages are particle/pore size distribution, electronic resistance, rheology, viscosity, viscoelasticity, gelation, time dependence, shear recovery, density, and surface tension. For optimum mixing stage, key performance IPVs involved in the

mixing stage would present a narrow distribution to indicate reliability and repro-
ducibility [17]. It should be noted that through correlating slurry IPVs with the
final cell properties, electrode with desired properties could be engineered. For this
purpose, it is therefore desirable to measure as many IPVs as possible.

Coating and drying: Coating and drying are another two key stages of electrode
manufacturing. To better explore these two stages, as many properties as possible
should be recorded, among which mass loading/thickness, active-to-inactive ratio,
solvent quantity, solid content, drying rate, temperature, gas flow rate, pressure, radiation intensity, coating techniques, and binder types. The corresponding IPVs here
are solvent evaporation rate, film thickness, components distribution or segregation,
surface temperature profile, morphology, elasticity, conductivity, and 3D microstructure. To obtain a target coating result, equipment settings such as comma bar gap,
coating ratio, web speed, and bumper bar gap should be carefully tuned. It should be
noted that the binder type, solid content, and web speed are not only the dominated
parameters but also on weight/thickness, temperature and air pressure in the oven.

Calendaring: Porosity degree would be controlled by the applied pressure and should
be optimized to offer good wettability [18], adhesion, and electrode conductivity
properties. Calendaring would strongly depend on the previous stages. For example,
it could compact a coating re-establishing the conductive paths, but when initial
porosity becomes too low, defects would be introduced into the network. A study
concerning the influence of calendaring pressure and speed on the formed electrode
has been researched. The investigations cover the initial electrode porosity, composition on $LiNi_{1/3}Mn_{1/3}Co_{1/3}O_2$ cathodes, electrolyte effective conductivity, fraction
of current collector covered, fraction of active material surface in contact with the
electrolyte and final electrode energy density. A conclusion is made to illustrate the
complex interplay of calendaring IPP and cell performance and the trade-offs that
need to be considered [19]. For instance, while the porosity after calendaring needs
to be minimized, the fraction of the current collector covered by the binder needs to
be maximized.

Electrode cutting, cell assembly, and electrolyte filling: There are two options
available for electrode cutting, both with benefits and drawbacks: die-cut and laser-
beam cutting [20, 21]. For the die-cut, the delamination at anode edges and the
bending of the current collector would be caused. For the laser-beam cutting, it could
produce less damage on anodes and performs better in a cell. However, the aluminium
spatter can promote dendrite growth for cathode cases. For battery cell assembly,
electrode–separator assembly becomes a source of variation. It should be noted that
the electrode misalignment would significantly affect the charge/discharge capacity
and ageing process. This could first reduce the anode–cathode overlapping area, and
secondly deposit Li-ion at the edges, initiating dendrite growth [22, 23]. Separators
are chosen based on their characteristics such as thickness, weight (which affect the
gravimetric and volumetric performance of a cell), ionic resistance (movement of
Li-ions between electrodes), and wettability and porosity (influencing electrolyte
quantity and cell life) [24, 25]. The choice of separator would also influence soaking

time and electrolyte quantity. The electrolyte is the ionic conductor between active material of electrode to ensure ion exchange. The electrolyte filling and wetting are final steps in the cell manufacturing process which plays the dominant role in the final battery performance. The unwetted or poorly wetted electrodes will not engage in the electrochemical reactions or cause increased impedance and dendrite formation that would lead to a short circuit.

Based upon the aforementioned discussions, battery manufacturing is extremely complicated with numerous IPPs and IPVs involved, which are also corresponding to the intermediate product or final battery product performance. In this context, generating a smarter battery manufacturing line with low waste, efficient, high quality, reproducible, and cost-effective stages is extremely challenging. Data science tools here can first benefit the understanding of these IPPs and IPVs involved in the battery manufacturing line. Furthermore, according to a well-trained data science model, numerous parameters could be simultaneously considered to forecast battery performance without making cells. This would be an efficient and cost-effective way to guide the future battery industry for managing battery manufacturing stages.

2.3 Li-Ion Battery Operation

Li-ion battery operation is the second stage of its lifespan. After production, all Li-ion batteries would inevitably age with time, losing their capacity to store charge and deliver it efficiently. Once a battery capacity falls below 80%, both battery range and performance would fall dramatically. Benefit from the rapid expansion of sustainable EV, Li-ion battery has been widely used in the transportation electrification field among all existing energy storage solutions [2]. In recent years, the frequent accidents of EVs have pushed EV to the subject of public opinion and also put forward high requirements and challenges for battery operation management [3]. In this context, battery operation stage here represents the period that battery capacity falls from 100 to 80%. As one of the most important components within EV, the operation management of Li-ion battery is crucial to the industrialization and marketization of EVs. Therefore, developing advanced and effective data science-based strategies for Li-ion battery operation management has become a hot research topic.

2.3.1 Battery Operation Fundamental

Figure 2.5 illustrates the functional structure diagram of typical battery operation management. To achieve effective battery operation management, various sensors, actuators, communication modules, signal lines, and operation strategies are required. The main task here is to guarantee battery safety and reliability during its operation stage, while extracting key information of battery internal states and lifetime for charging control and energy management. Furthermore, the abnormal cases

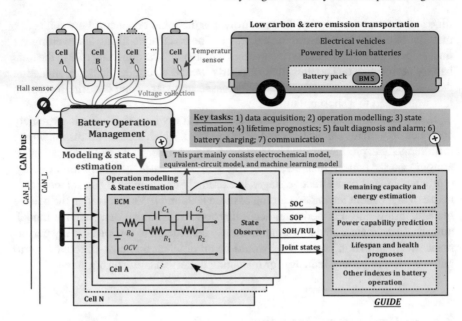

Fig. 2.5 Functional structure diagram of battery operation management

and battery faults should be detected or predicted while appropriate interventions need to be taken.

The technical challenges restricting the development of battery operation management strategies mainly include the following three aspects: (1) Li-ion battery has highly nonlinear behaviour during its operation, with multi-spatial scale (such as electrical dynamics, and thermal dynamics, etc.) and multi-time scale, making its operation behaviours difficult to be accurately captured; (2) numerous battery internal states are difficult to be obtained through direct measurement solution and would be easily influenced by environmental conditions such as temperature and noise. The upsizing of power Li-ion battery would decrease the representativeness of measured signals, and reduce the predictability of battery states, making it become difficult to effectively estimate battery internal states; (3) as battery health plays a pivotal role in determining battery safety and efficiency but it would inevitably deteriorate with time, a key but challenging issue for efficient battery operation lies in the battery ageing prognostics and health management; (4) various faults within Li-ion batteries would potentially cause their performance degradation and severe safety problems. Developing advanced fault diagnosis strategies is becoming increasingly critical for the efficient and safe operation of Li-ion battery; and (5) charging strategy directly affects the battery charging speed, energy conversion efficiency, temperature variation, and battery degradation, which becomes difficult to be controlled for well balancing these important but conflicting charging objectives. In this context, advanced battery operation management solutions need to be designed for solving the aforementioned issues.

2.3.2 Key Tasks of Battery Operation Management

To ensure battery safety and performance during its operation period when capacity degrades from 100 to 80%, some key tasks of battery operation management include operation modelling, state estimation, lifetime/ageing prognostics, fault diagnosis, and battery charging are explored in this book, as illustrated in Fig. 2.6.

(1) **Operation modelling**

Battery is a typical electrochemical component with strong-coupled electrical-thermal behaviours. Establishing a high-fidelity battery operation model plays a pivotal role in the further design of operation management algorithms. The performance in terms of accuracy, flexibility, and robustness of an operation model would directly influence the related battery states estimation, lifetime prognostics, fault diagnosis, and charging strategies. The battery electrical, thermal, ageing, and coupled characteristics could determine the main form of battery operation models, as illustrated in Fig. 2.7.

Battery electrical model is able to describe battery electrical behaviours such as voltage response during charging and discharging, a basic but key element during battery operations. As temperature would significantly affect battery performance and safety during its operation, establishing a reliable thermal model to capture battery thermal dynamics is also critical. Furthermore, as battery capacity and power would degrade gradually during operation, an effective ageing model could help to monitor the dynamics of battery degradation. From a higher systematic perspective, temperature would affect

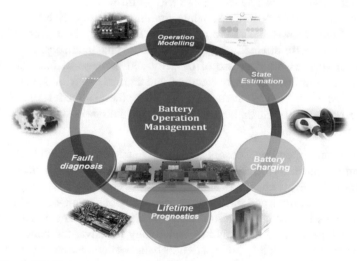

Fig. 2.6 Key tasks of battery operation management

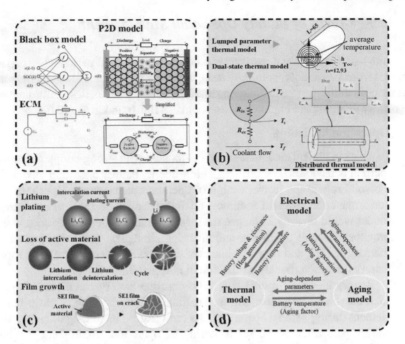

Fig. 2.7 Schematic illustration of battery operation modelling. **a** electrical model, **b** thermal model, **c** ageing model, and **d** coupled model, reprinted from [26], with permission from Elsevier

battery electrical behaviours and degradation rate. For example, the growth of battery solid electrolyte interphase (SEI) film would become quicker under high-temperature operations, while lithium deposition is easier to be triggered under low-temperature operations. The battery heat generation would be also directly affected by battery electrical characteristics. Battery degradation such as the decrease of capacity and increase of internal resistance would also influence battery electrical and thermal behaviours. In this context, coupled battery models such as battery electrothermal model are also required to capture multi-domain battery behaviours simultaneously.

(2) **State estimation**

For battery operation, sensors can directly and conveniently measure battery terminal current, voltage, and surface temperature. However, owing to complex electrochemical behaviours and multi-physics coupling, just using the signals of voltage, current, and surface temperature obviously cannot reflect all battery dynamics during its operation. How to effectively estimate battery internal states constitutes a key enabling technology for advanced battery operation management. Credible information of battery state of charge (SoC), state of power (SoP), and state of health (SoH) is a prerequisite for effective charging, power, and health management of batteries, which require to be accurately estimated online, as illustrated in Fig. 2.8.

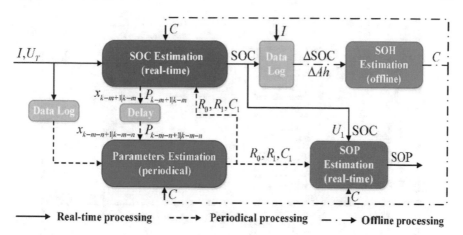

Fig. 2.8 A framework for battery SoC, SoP, and SoH estimations, reprinted from [27], with permission from Elsevier

(3) **Lifetime/ageing prognostics**

Li-ion battery performance would deteriorate with time due to the degradation of its electrochemical constituent, leading to battery capacity and power fade [2]. This is named battery ageing and is a consequence of multiple coupled mechanisms affected by various factors such as battery chemistry, environmental and operating conditions. The time period when a battery fails to satisfy the demands of energy or power during its operation is generally defined as battery lifetime. To ensure the safety and satisfactory performance of the battery despite degradation, battery lifetime prognostic tools are required.

The prognostics of Li-ion battery lifetime/ageing concerns the energy or power degradation of battery in the future and predict how soon the performance of battery would be unsatisfactory [7]. Figure 2.9 illustrates a battery lifetime prognostics framework from offline model development to online prediction. In general, both the current and historical battery ageing information is required to perform effective battery lifetime prediction. The current battery health information could be obtained from battery SoH estimator, while the historical battery ageing information is recorded by computer. Then the future health state and lifetime of battery under a certain operation case could be predicted. With the predicted battery lifetime information, battery users can know the battery service life and schedule any maintenance or replacement in advance.

(4) **Fault diagnosis**

Without suitable battery fault diagnosis, a minor fault would also eventually result in severe damage to battery [29]. The importance of battery fault diagnosis has been demonstrated in many severe battery incidents [30, 31]. During battery operation, numerous fault modes exit and fault mechanisms

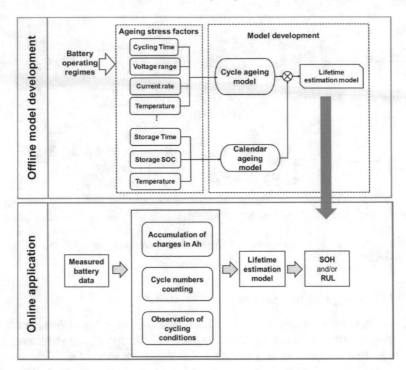

Fig. 2.9 Battery lifetime/ageing prognostics from offline model development to online prediction, reprinted from [28], with permission from Elsevier

would be generally complicated. From a control perspective, the operation fault modes mainly include battery fault, sensor fault, and actuator fault, as illustrated in Fig. 2.10 [29]. For the battery fault, overcharging, overdischarging, overheating, external and internal short circuits, electrolyte leakage, swelling, accelerated degradation, and thermal runaway become the most critical ones during battery operation. These faults could be also intertwined. For example, overcharging and overdischarging would result in different undesirable side reactions within batteries, further leading to accelerated battery degradation. The side reaction and gas generated by the chain reaction of a thermal runaway could also eventually cause battery swelling. Such battery swelling, together with mechanical damage, could, in turn, result in electrolyte leakage.

For the fault of an internal short circuit, it is usually caused by the separator failure from the manufacturing defects, overheating, and mechanical collisions. Fortunately, the Joule heat generated by internal short circuit will develop into a thermal runaway only in the case that the equivalent resistance reaches a very low level. It should be known that abnormal heat generation will occur in different conditions including the side reaction during overcharging or overdischarging, external and internal short circuits, as well as the contact loss of cell connectors. This would significantly increase battery temperature. Overheating

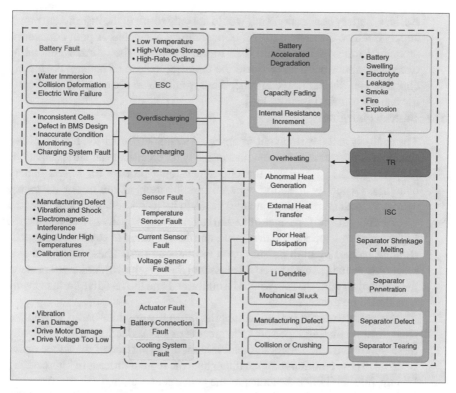

Fig. 2.10 Faults of Li-ion battery operation, reprinted from [29], with permission from IEEE

becomes the direct cause of battery thermal runaway and would be facilitated by a chain reaction of thermal runaway [32], leading to the vicious positive feedback cycle.

Apart from battery fault itself, sensor fault would also lead to severe issues for Li-ion battery operation as all the feedback-based battery operation management strategies significantly rely on the sensor measurement [33]. Sensor faults during battery operation can be mainly divided into the voltage sensor fault, current sensor fault, as well as temperature sensor fault. The current sensor fault would highly influence the accuracy of battery state estimation. Li-ion battery needs to be operated under safe voltage and temperature ranges. Operating batteries outside these ranges could lead to a decrease in battery performance and even cause severe accidents.

For the actuator fault, it could play a more direct effect on the control system performance than do battery fault as well as sensor fault. Potential actuator faults during battery operation include the terminal connector fault, cooling system fault, controller area network bus fault, high-voltage contactor fault, and fuse fault. If a cooling system fails, it is difficult to maintain battery within a suitable temperature operation range, which would even trigger thermal runaway.

Besides, battery connection fault would not only result in the insufficient power supply but also the increased risk of operational accidents. A poor connection among Li-ion batteries would lead to the resistance rise, further generating excessive, abnormal heat to cause the temperature rise of battery. As battery charging and discharging continue, there would be an arc or spark, leading to the battery terminal melting [34].

(5) **Battery charging**

The analysis and development of charging strategies for Li-ion battery play a vital role in operation management especially for electrical vehicle applications [35]. Figure 2.11 illustrates the one-dimensional process for Li-ion battery charging. Generally speaking, when a Li-ion battery is run out of its energy source, it must stop discharging and needs to be recharged. It should be noted that the charging performance of Li-ion batteries is limited by two key elements: lithium plating happened on the anode, and the oxidation of electrolyte aqueous solution caused by high potentials on the cathode [36]. These undesired side reactions result in some irreversible losses of the cyclable lithium. Furthermore, the electrolyte components would be consumed and the resistive surface layers would also be thickened.

The anode material for Li-ion batteries is usually graphite-based. On the one hand, lithium plating represents the process of Li^+ reduction, which is finally dissolved in the electrolyte, to metal lithium at the surface of the anode active material. These reactions happen rather than the regular process of intercalating lithium into the lattice structure of the active material, and the reactions can step from the limitations for charging transfer or lithium solid diffusion [37]. When the anode potential reduces to a value lower than the standard potential of Li^+/Li, lithium plating would occur. Graphite-based anodes are very active and easy to cause lithium plating due to their low equilibrium potential, especially at the high level of SoC. The trend of lithium plating becomes greater under

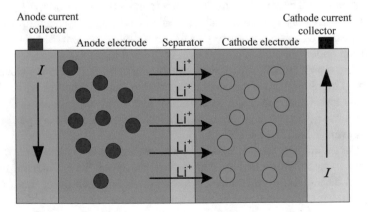

Fig. 2.11 One-dimensional process for Li-ion battery charging

the conditions of higher charging current, higher SoC, and lower temperature. Furthermore, the mechanical stress and volume changes would happen when the lithium is intercalated into the graphite anode, which can result in battery ageing. As illustrated in [38], an aged battery can be more impressionable to lithium plating. In a word, the intercalation kinetics at the anode mainly limits the level of charging currents for graphite-based Li-ion batteries.

On the other hand, the oxidation of electrolyte solvents which occurs at high cathode potentials would further limit the level of charging voltage. When the cathode material is fully delithiated, the irreversible damage to the crystallographic structure of the cathode can be caused by overcharging a Li-ion battery to the impertinent voltage level, which results in further oxidative side reactions. Moreover, these side reactions can entail gas evolution, overpressure inside the battery cell, opening of the cell's safety vest, and leakage of electrolytes. Fire or explosion of the battery would also occur due to impertinent voltage since the organic electrolytes of Li-ion battery are highly flammable. All in all, the maximum battery charging voltage level is essential for the safe operation of Li-ion batteries charging.

In general, charging strategies based on the external characterization of the battery and the electrochemical characteristics during charging are vital to enhancing battery operation performance. Theoretically, the charging time can be shortened by increasing the charging current amplitude, which, however, will cause large electrochemical reaction stresses, and thus accelerate battery degradation and reduce its lifespan.

2.4 Li-Ion Battery Reutilization

Li-ion battery has become the mainstream power source in EVs. When the battery capacity decays to around 80% of its initial capacity, it needs to be retired from EVs to ensure safety. With the explosive development of EVs, more and more batteries are retired from EVs. How to safely and reliably dispose of these retired batteries has become a global problem. These retired batteries have considerable economic and environmental value. On the one hand, although these batteries cannot be used in EVs, they can be used in occasions with lower safety requirements than EVs, such as energy storage power and standby power supplies. This secondary utilization is called echelon utilization. On the other hand, Li-ion batteries have rich resource properties, which contain a large number of valuable metals. If these valuable metals are recycled, the mining of raw ore can be reduced, which has great economic and environmental value. It can be seen that the retired Li-ion batteries can be reused, and the reutilization includes echelon utilization and material recycling.

Figure 2.12 shows the technical route of secondary utilization of Li-ion batteries. The battery pack retired from EVs has two technical routes: (a) If the performance and consistency of the battery pack are good, the battery can be repaired and reused through some technical means. The common methods include replacing

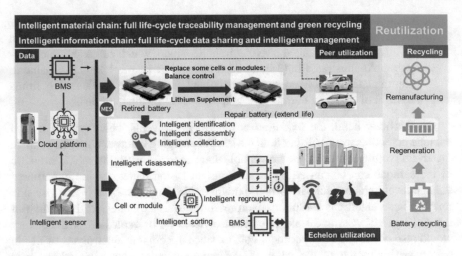

Fig. 2.12 Schematic diagram of Li-ion battery reutilization

some modules or cells, balancing, lithium supplement, etc. (b) For the battery packs with great differences in battery health status, echelon utilization is required. The process of echelon utilization is divided into three parts: battery disassembly, battery sorting, and battery regrouping. They are described as follows:

(1) Battery disassembly. The pack-level utilization is not a good scheme. Generally, the battery pack needs to be disassembled to the module or cell level. At present, most disassembly is manual, which has high labour costs and chemical hazards. Automatic disassembly by robot is an important direction.

(2) Battery sorting. Its purpose is to evaluate the residual value of the battery. The historical data of battery is a valuable resource for the battery performance evaluation. If the battery has available historical data, the state of the battery can be easily obtained. On the contrary, the battery needs to be tested to evaluate the performance of the battery, which is time-consuming and energy-consuming. Therefore, it is very important to build a Li-ion battery lifecycle data platform.

(3) Battery regrouping. At this stage, the sorted cells or modules are regrouped based on the echelon utilization scenario. There are many parameters to characterize the performance of the battery. How to regroup these batteries is a technical problem. In addition, the regrouped battery needs to be managed by the battery management system for secondary life. The safety of battery during secondary utilization is a key issue to be concerned about.

When the battery is retired for the second time from the echelon utilization scenario, it will enter the battery recycling occasion. At present, there are many schemes for battery recycling, among which pyrometallurgy and hydrometallurgy are more common. Developing green recycling with low energy consumption, low pollution, and all components is an important direction. There are two technical routes in the material recycling stage: (a) direct regeneration. The components of waste batteries are directly chemically and physically treated to restore their performance. This method has a good economy, but the technology is still mature. (b) Battery remanufacturing. Waste batteries are recycled as materials for the manufacture of new batteries. In short, battery reutilization has two chains: material and information chains. The development trend of the material chain is full lifecycle traceability management and green recycling, and the development trend of information chain is full lifecycle data sharing and intelligent management.

2.5   Summary

This chapter introduces the key stages for battery full-lifespan management. It first offers the concept and gives a comprehensive framework about the full-lifespan of Li-ion battery, which can be mainly divided into three stages including battery manufacturing, battery operation, and battery reutilization. Then the management tasks of each stage arc introduced in detail. For battery manufacturing which can be divided into electrode manufacturing and cell manufacturing, as it involves numerous manufacturing parameters and variables with strong-coupled relations, the main management task here is to derive reliable data science strategy for decoupling correlations among various manufacturing variables and further exploring the impact variables towards resulting manufactured battery performance. For battery operation that refers to EV applications, some key tasks including battery operation modelling, state estimation, lifetime/ageing prognostics, fault diagnosis, and charging need to be explored to ensure battery safety and performance when its capacity degrades from 100 to 80%, the battery is first used for echelon utilization to make full use of its residual capacity. When the capacity decays below around 40% of its nominal capacity, the battery will be retired for the second time for material recycling, and the recycled materials will be used for battery remanufacturing.

These mentioned management tasks indicate that battery full-lifespan is actually a close-loop progress while each task from different stages should be well managed to make full use of battery. As battery full-lifespan also generates a great deal of data to reflect different battery dynamics, effective data science-based solutions are promising ways to handle these tasks and ensure the high-performance of battery.

References

1. Blomgren GE (2016) The development and future of lithium ion batteries. J Electrochem Soc 164(1):A5019
2. Kwade A, Haselrieder W, Leithoff R, Modlinger A, Dietrich F, Droeder K (2018) Current status and challenges for automotive battery production technologies. Nat Energy 3(4):290–300
3. Boston Consulting Group Report. The future of battery production for electric vehicles. https://www.bcg.com/publications/2018/future-battery-production-electric-vehicles

4. Bresser D, Buchholz D, Moretti A, Varzi A, Passerini S (2018) Alternative binders for sustainable electrochemical energy storage—the transition to aqueous electrode processing and bio-derived polymers. Energy Environ Sci 11(11):3096–3127
5. Lenze G, Bockholt H, Schilcher C, Froböse L, Jansen D, Krewer U, Kwade A (2018) Impacts of variations in manufacturing parameters on performance of lithium-ion-batteries. J Electrochem Soc 165(2):A314
6. Mohanty D, Hockaday E, Li J, Hensley D, Daniel C, Wood III D (2016) Effect of electrode manufacturing defects on electrochemical performance of lithium-ion batteries: cognizance of the battery failure sources. J Power Sources 312:70–79
7. Baunach M, Jaiser S, Schmelzle S, Nirschl H, Scharfer P, Schabel W (2016) Delamination behavior of lithium-ion battery anodes: influence of drying temperature during electrode processing. Drying Technol 34(4):462–473
8. Gutoff EB, Cohen WD (1996) R&D needs in the drying of coatings. Drying Technol 14(6):1315–1328
9. Stein IV M, Mistry A, Mukherjee PP (2017) Mechanistic understanding of the role of evaporation in electrode processing. J Electrochem Soc 164(7):A1616
10. Thorat IV, Stephenson DE, Zacharias NA, Zaghib K, Harb JN, Wheeler DR (2009) Quantifying tortuosity in porous Li-ion battery materials. J Power Sources 188(2):592–600
11. Schilling A, Gümbel P, Möller M, Kalkan F, Dietrich F, Dröder K (2018) X-ray based visualization of the electrolyte filling process of lithium ion batteries. J Electrochem Soc 166(3):A5163
12. Schilling A, Gabriel F, Dietrich F, Dröder K (2019) Design of an automated system to accelerate the electrolyte distribution in lithium-ion batteries. Int J Mech Eng Robot Res 8(1)
13. Qu D, Fundamental principals of battery design: porous electrodes. In: Proceedings of American Institute of Physics Conference (AIP), TU Bergakademie, Germany, Freiberg, 2014, pp 14–25
14. Zhou Y, Su M, Yu X, Zhang Y, Wang J-G, Ren X, Cao R, Xu W, Baer DR, Du Y (2020) Real-time mass spectrometric characterization of the solid–electrolyte interphase of a lithium-ion battery. Nat Nanotechnol 15(3):224–230
15. Zubi G, Dufo-López R, Carvalho M, Pasaoglu G (2018) The lithium-ion battery: state of the art and future perspectives. Renew Sustain Energy Rev 89:292–308
16. Mishra A, Mehta A, Basu S, Malode SJ, Shetti NP, Shukla SS, Nadagouda MN, Aminabhavi TM (2018) Electrode materials for lithium-ion batteries. Mater Sci Technol 1(2):182–187
17. Hoffmann L, Grathwol J-K, Haselrieder W, Leithoff R, Jansen T, Dilger K, Dröder K, Kwade A, Kurrat M (2020) Capacity distribution of large lithium-ion battery pouch cells in context with pilot production processes. Energy Technol 8(2):1900196
18. Sheng Y, Fell CR, Son YK, Metz BM, Jiang J, Church BC (2014) Effect of calendering on electrode wettability in lithium-ion batteries. Front Energy Res 2:56
19. Duquesnoy M, Lombardo T, Chouchane M, Primo E, Franco AA (2020) Accelerating battery manufacturing optimization by combining experiments. In silico electrodes generation and machine learning
20. Pfleging W (2018) A review of laser electrode processing for development and manufacturing of lithium-ion batteries. Nanophotonics 7(3):549–573
21. Jansen T, Kandula MW, Hartwig S, Hoffmann L, Haselrieder W, Dilger K (2019) Influence of laser-generated cutting edges on the electrical performance of large lithium-ion pouch cells. Batteries 5(4):73
22. Heins TP, Leithoff R, Schlüter N, Schröder U, Dröder K (2020) Impedance spectroscopic investigation of the impact of erroneous cell assembly on the aging of lithium-ion batteries. Energy Technol 8(2):1900288
23. Leithoff R, Fröhlich A, Dröder K (2020) Investigation of the influence of deposition accuracy of electrodes on the electrochemical properties of lithium-ion batteries. Energy Technol 8(2):1900129
24. Francis CF, Kyratzis IL, Best AS (2020) Lithium-Ion battery separators for ionic-liquid electrolytes: a review. Adv Mater 32(18):1904205

25. Weber CJ, Geiger S, Falusi S, Roth M (2014) Material review of Li ion battery separators. In: Proceedings of American Institute of Physics Conference (AIP), TU Bergakademie, Germany, Freiberg, 2014, pp 66–81
26. Dai H, Jiang B, Hu X, Lin X, Wei X, Pecht M (2021) Advanced battery management strategies for a sustainable energy future: multilayer design concepts and research trends. Renew Sustain Energy Rev 138:110480
27. Hu X, Jiang H, Feng F, Liu B (2020) An enhanced multi-state estimation hierarchy for advanced lithium-ion battery management. Appl Energy 257:114019
28. Li Y, Liu K, Foley AM, Zülke A, Berecibar M, Nanini-Maury E, Van Mierlo J, Hoster HE (2019) Data-driven health estimation and lifetime prediction of lithium-ion batteries: a review. Renew Sustain Energy Rev 113:109254
29. Hu X, Zhang K, Liu K, Lin X, Dey S, Onori S (2020) Advanced fault diagnosis for lithium-ion battery systems: a review of fault mechanisms, fault features, and diagnosis procedures. IEEE Ind Electron Mag 14(3):65–91
30. Abada S, Marlair G, Lecocq A, Petit M, Sauvant-Moynot V, Huet F (2016) Safety focused modeling of lithium-ion batteries: a review. J Power Sources 306:178–192
31. Kleinman Z. BBC News. Samsung recalls Note 7 flagship over explosive batteries. https://www.bbc.com/news/business-37253742
32. Wang Q, Ping P, Zhao X, Chu G, Sun J, Chen C (2012) Thermal runaway caused fire and explosion of lithium ion battery. J Power Sources 208:210–224
33. Dey S, Mohon S, Pisu P, Ayalew B (2016) Sensor fault detection, isolation, and estimation in lithium-ion batteries. IEEE Trans Control Syst Technol 24(6):2141–2149
34. Ma M, Wang Y, Duan Q, Wu T, Sun J, Wang Q (2018) Fault detection of the connection of lithium-ion power batteries in series for electric vehicles based on statistical analysis. Energy 164:745–756
35. Tomaszewska A, Chu Z, Feng X, O'kane S, Liu X, Chen J, Ji C, Endler E, Li R, Liu L (2019) Lithium-ion battery fast charging: a review. ETransportation 1:100011
36. Aurbach D (2000) Review of selected electrode–solution interactions which determine the performance of Li and Li ion batteries. J Power Sources 89(2):206–218
37. Ramadesigan V, Boovaragavan V, Pirkle JC Jr, Subramanian VR (2010) Efficient reformulation of solid-phase diffusion in physics-based lithium-ion battery models. J Electrochem Soc 157(7):A854
38. Dubarry M, Liaw BY, Chen M-S, Chyan S-S, Han K-C, Sie W-T, Wu S-H (2011) Identifying battery aging mechanisms in large format Li ion cells. J Power Sources 196(7):3420–3425

Chapter 3
Data Science-Based Battery Manufacturing Management

This chapter focuses on the data science technologies for battery manufacturing management, which is a key process in the early lifespan of battery. As a complicated and long process, the battery manufacturing line generally consists of numerous intermediate stages involving strongly coupled interdependency, which would directly determine the performance of the manufactured battery. In this context, the in-depth exploration and management of different manufacturing parameters, variables, their correlation as well as effect towards the resulted property of manufactured intermediate products or final battery performance is crucial but still remains a difficult challenge. Recent advancements in data-driven analytic and related machine learning strategies raised interest in data science methods to perform effective and reasonable management of battery manufacturing.

To give a systematic description of how to develop data science methods to benefit battery manufacturing management, an introduction is first given to dividing battery manufacturing into two main parts including battery electrode manufacturing and battery cell manufacturing. Then the data science framework and related machine learning tools for battery manufacturing management are described in detail. In addition, for both battery electrode manufacturing and cell manufacturing, two case studies of deriving proper data science methods to benefit their management are presented and discussed, respectively. This chapter would inform insights into the feasible data science methods with interpretability for the effective classification of battery product quality, prediction of manufactured battery performance, and sensitivity analyses of different manufacturing parameters of interest, further promoting smart battery manufacturing management.

© The Author(s) 2022
K. Liu et al., *Data Science-Based Full-Lifespan Management of Lithium-Ion Battery*, Green Energy and Technology,
https://doi.org/10.1007/978-3-031-01340-9_3

3.1   Overview of Battery Manufacturing

As the first key stage of battery full-lifespan, the performance of Li-ion batteries would be directly and highly influenced by their manufacturing process, which significantly hinders the improvement of battery technologies. Suitable management of battery manufacturing plays a pivotal role in developing clean and efficient battery-based energy storage systems, which is also a key factor to secure tangible economic payback and to improve the efficiency of large-scale clean energy applications [1]. In this context, efforts are urgently required to fully understand the specific production stages, the intermediate products as well as the production parameters within a battery manufacturing chain [2].

Unfortunately, battery manufacturing chain contains numerous intermediate stages with lots of intermediate products, parameters, and impact variables [1]. Figure 3.1 illustrates the general processes of Li-ion battery manufacturing, which can be mainly divided into three parts including the battery material preparation, electrode manufacturing, and battery cell manufacturing. In general, for material preparation part, components to produce battery such as the active material, electrode additive, polymeric binder would be prepared based on the requirements of different types of Li-ion battery. These components and related formulations must be carefully selected as they significantly affect the manufactured electrode qualities such as electronic conductivity and thickness. For electrode manufacturing part, after preparing suitable component materials, these materials will be mixed to generate slurry. Then the slurry will be coated onto the surface of metal foil, followed by a drying stage in an oven with the preset temperature and then a calendaring stage to evaporate the residual solvent. For battery cell manufacturing part, the manufactured electrode would be cut into various sizes and then assembled. Then the electrolyte

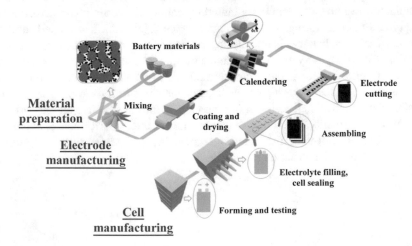

Fig. 3.1 Schematic diagram of typical Li-ion battery production processes

will be filled and the battery cell will be sealed, followed by the forming and testing steps to finish cell manufacturing.

It is worth mentioning that due to the complexity and strong-coupled interdependencies within each battery manufacturing part, the multiple correlations among feature variables of intermediate products and control parameters are still difficult to model. As the whole battery manufacturing chain consists of a number of chemical, mechanical as well as electrical operations and would generate over 600 influencing parameters or variables, the analysis of feature variables in battery manufacturing requires deep expert experiences and specialized equipment, which still mainly relies on the trial and error [3]. Therefore, in order to achieve smarter and cleaner battery production, advanced data science strategies to better quantify feature variables and select key feature items for predicting battery electrode properties are urgently needed.

With the rapid development of artificial intelligence and machine learning technologies, data-driven strategies have become popular in the field of battery management. A range of data-driven approaches have been designed for battery applications [4, 5]. Overall, by designing suitable data-driven models, it is expected that smarter and more efficient management of Li-ion batteries can be achieved. However, these researches primarily focus on the battery macroscopic performance without taking the battery intermediate properties in the production process into account. It should be noted that the battery production chain also generates a large amount of data and plays a more direct role in determining the battery performance, designing an effective data-driven approach to quantify and predict battery intermediate features is therefore also crucial for boosting the development of cleaner production [6].

In contrast to battery management where fruitful solutions are available, fewer reports have been found so far on using advanced machine learning technologies to improve battery production [7]. Among limited literature on battery productions (e.g. monitoring, adjustment, and control), developing a proper data-driven-based model to analyse feature variables and predict intermediate product properties is a hotspot. For instance, a data-driven approach was proposed in [8] to analyse the failure modes and parametric effects, which contributes to the improvement of battery production chain control. Based on the cross-industry standard process (CRISP), Schnell et al. [9] designed a linear model and a neural network model to identify the process dependencies and forecast the battery production properties. Turetskyy et al. [3] utilized the decision tree techniques to analyse feature importance and forecast the maximum capacity of battery. A multi-variate data-driven model was designed in [10] to discover proper quality gates for predicting the manufactured battery properties. Based upon the statistical analyses of fluctuations in battery production, the influence of these fluctuations on manufactured battery capacity is evaluated in [11]. In [12], several 2D graphs produced by three conventional machine learning classification models are used to analyse the dependencies of battery production features.

Based upon the aforementioned works on the data-driven modelling of battery production, the main research focuses of data science-based battery manufacturing management can be divided into two parts including data collection as well as process analysis and property prediction, as illustrated in Fig. 3.2.

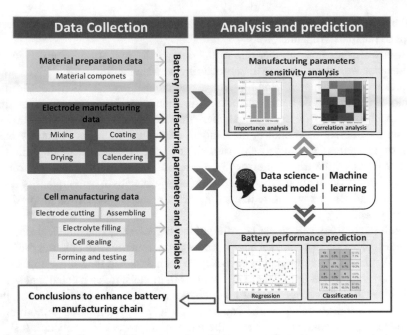

Fig. 3.2 Schematic diagram of data science-based battery manufacturing management

Data collection: Data collection is the first main stage for data science-based management of battery manufacturing. This stage would focus on the exploration of battery manufacturing chain, the collection and the management of original battery manufacturing data into manual data source or automatic data source. The manual data source usually represents the intermediate manufacturing data that need to be collected into database manually without an available interface. In contrast, for the automatic data collection, with the booming digitization equipment for battery manufacturing, the data could be automatically collected into database. Obviously, automatic data collection is capable of saving time and effort, and also guarantee the reliability of collected battery manufacturing data. In this context, some advanced digitization systems such as traceability systems [13, 14] are worth developing, further popularizing the automatic data collection for the data science-based management of battery manufacturing.

Process analysis and property prediction: Here the intermediate process analysis mainly focuses on understanding the nature of the battery manufacturing process through analysing the importance and correlation of manufacturing parameters based on suitable data science tools. After collecting battery manufacturing data, these data would be well preprocessed to generate suitable data matrix for exploration [15]. Then the state-of-the-art machine learning tools would be used to derive proper data science models to analyse the sensitivity of interested parameter items within battery manufacturing chain [16, 17]. In general, both process parameters and intermediate product variables of each battery manufacturing step would be set as the explored targets. Due to battery manufacturing actually belonging to a step-by-step operational line, the outputs from the previous manufacturing stage could be utilized as the inputs of the next manufacturing stage for sensitivity analysis. After visualizing the analysed results by heat map or human-interface graphs, battery manufacturers can get useful conclusions and new ideas to readjust manufacturing parameters and optimize the management of battery manufacturing. On the other hand, battery cell properties prediction mainly focuses on developing suitable data science models to perform effective prognostics for both battery product performance and other manufacturing targets at the battery early manufacturing stage. Here two properties need to be considered. The first one is the manufactured battery cell performance such as its capacity [18], power or energy densities, service life. Another one is battery manufacturing goals such as the reduced cost and energy loss, the avoided manufacturing waste, and the increased battery yield.

3.2 Data Science Application of Battery Manufacturing Management

For the data science applications of battery manufacturing management, there are two main crucial things should be carefully considered. One is the utilized framework of designing data science-based method to perform analysis or predictions within battery manufacturing chain and another is the machine learning solutions to design related data science model.

3.2.1 Data Science Framework for Battery Manufacturing Management

First, the data science-based framework would significantly influence the efficiency and results of battery manufacturing analyses. In order to efficiently analyse manufacturing data from the whole battery manufacturing chain including the material preparations, electrode and cell manufacturing, a classical data science-based framework named the cross-industry standard process for data-mining (CRISP-DM) is widely utilized [3].

Figure 3.3 details the six steps of CRISP-DM framework, which includes the business understanding step, data understanding step, data preparation step, modelling step, evaluation step, and deployment step. To be specific, for the business understanding step, after well exploring the battery manufacturing chain of interest, the requirement, target and restriction of data science activities could be first defined from the business perspective. Then in the data understanding step, through using some direct data analytic and visualization methods such as heat map and scatter plot, the raw data from battery manufacturing chain is initially investigated for further

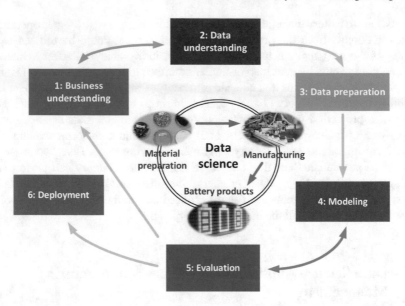

Fig. 3.3 CRISP-DM-based data science framework of battery manufacturing management

deriving newly required data and quality. Besides, the obtained observations from data understanding stage could be adopted to readjust back data science target of the business understanding stage. Then the next data preparation step focuses on the preparation of suitable dataset extracted from battery manufacturing data for modelling activities [19]. This stage would be significantly time-consuming but also very important as the quality of battery manufacturing data would highly affect the performance of model training. In this context, several data preprocessing strategies such as curation, formatting, integration, and standardization would be adopted for generating high-quality manufacturing data. After that, suitable machine learning tools need to be carefully selected for deriving data science models based on different data analysis goals. In this stage, data science model would be generally optimized by readjusting its input form and updating both model structure and parameters set to achieve satisfactory analyses or predictions for battery manufacturing. Then in the evaluation stage, the modelling results especially for the sensitivity analyses of battery manufacturing variables and the predicted cell properties would be further explored with the predefined data science target, which would result in the new data science goals of battery manufacturing. Here the obtained results as well as conclusions can be visualized directly with suitable human interpretable graphs. These information could help battery manufacturers to get useful information and to create new evaluation plans for analysing battery manufacturing process. Based upon these observations obtained from data science tools, the improved solution or further evaluation could be proposed to optimize the intermediate stages for achieving smarter battery manufacturing management.

3.2.2 Machine Learning Tool

Apart from data science framework, machine learning tool is another crucial element for achieving effective data science-based battery manufacturing management. As there are numerous existing machine learning tools, the key issue here becomes how to select a suitable one to provide reliable data analysis results for different data science applications [20]. The machine learning tools for data science-based battery manufacturing management generally require the information of process parameters or variables from battery manufacturing chain as the inputs to predict the properties of both intermediate products and final manufactured batteries.

To well reflect the nature of machine learning tools in the management of battery manufacturing, a classification form is utilized to divide these state-of-the-art machine learning tools into non-interpretable and interpretable ones. For non-interpretable tools, just the predicted point result is given without obtaining information of importance and correlations of manufacturing parameters of interest. Yet, a key aspect of data analyses in battery manufacturing chain is not only generating the predicted property value but also providing the sensitivity analyses of the investigated process parameters or variables. In light of this, the interpretable machine learning tools are gaining much more attention for the smart management of battery manufacturing.

It should be known that the selection of suitable machine learning tools for data science-based battery manufacturing management is a multifaceted problem, depending on the analysis target, the data size and the process objective. Table 3.1 systematically summarizes and compared different classical machine learning tools including the linear model, quadratic model, support vector machine (SVM), neural network (ANN), decision tree (DT), random forest (RF), and others to predict battery product properties and perform sensitivity analyses for the applications of battery manufacturing chain. It can be concluded that none of a machine learning tool could be regarded as the one-size-fits-all solution; instead, there is the inherent trade-off between computational effort and the performance of corresponding predictions as well as sensitivity analyses. Moreover, some machine learning tools could only provide a single prediction point with the regression or classification form. Ideally, however, the reliable sensitivity analyses for capturing the importance and correlation of manufacturing parameters, is becoming a rigid requirement because of a large number of process parameters or variables are existed in the battery manufacturing chain and some of them are strongly coupled with each other [21]. In addition, the interpretable machine learning tools such as DT and RF present reasonable abilities for providing parameter sensitivity analyses, giving predicted point results and quantifying the parameter importance as well as correlations within battery manufacturing chain. Due to this merit, machine learning tools with strong interpretability are preferable as these quantified sensitivity analysis results can well benefit battery manufacturers [22]. However, data science applications based on the interpretable machine learning tool for battery manufacturing management is still in their infancy. Most existing researchers test their interpretable data science models on their own

Table 3.1 Systematical comparison of typical machine learning tools for battery manufacturing management

ML tools	Interpretability	Advantages	Limitations
Linear model	Good	• Simple and compact • Easy to implement	• Low nonlinear capture ability • Poor generalization
Quadratic model	Good	• Easy to be built up • Easy to implement	• Low nonlinear capture ability • Difficult to analyse strongly coupled IPPS' correlations
SVM	None	• Nonparametric • Robust to outliers	• High computational burden
NN	Weak	• Good nonlinear fitting performance • High accuracy	• High computational burden • Easy to cause overfitting
DT	Good	• Flexible • Few parameters need to tune • Easy to explain	• Instability • Inadequate for predicting continuous variables
RF	Good	• Flexible • Few parameters to tune • Robust to outliers • Out-of-bag prediction	• High computational burden • Large memory requirements
Others	Depends on their nature	• Good manufacturing analysis performance • Good prediction accuracy	• Transferability is unclear • Robustness is unknown

manufacturing data, which will call into question the generalization of them in other battery manufacturing cases where the operation is significantly different. It is thus suggested to improve these interpretable machine learning tools by validating related data science model for more complex battery manufacturing cases. In addition, most works simply adopt these tools to analyse battery manufacturing data without an in-depth optimization of their performance. As the performance of machine learning-based data science model would be also significantly influenced by its structure and parameters. Advanced model optimization solution needs to be explored for further improving model performance in the field of battery manufacturing management.

3.3 Battery Electrode Manufacturing

3.3.1 Overview of Battery Electrode Manufacturing

Electrode is a key component of battery. The electrode manufacturing of Li-ion batteries belongs to a highly complex and long process that involves many disciplines such as electrical, chemical, and mechanical engineering. Figure 3.4 systematically illustrates several key steps for battery electrode manufacturing.

According to Fig. 3.4, after preparing proper materials, battery electrode manufacturing generally consists of several individual processes including mixing, coating, drying as well as calendaring. For the materials of manufacturing Li-ion battery electrode, active material, conductive additive, solvent and binder are generally required, as illustrated in Fig. 3.5. To be specific, $LiFePO_4$ (LFP) and $Li_4Ti_5O_{12}$ (LTO) are widely used as the active material due to the fact that LFP has the merits of being non-toxic and adaptable to large current rates and high temperatures, while LTO is capable of mitigating the irreversible formations of SEI as well as dendrite. Based on this, the manufactured battery can provide high power and long cycling behaviours. For electrode additives, conductive fillers including the

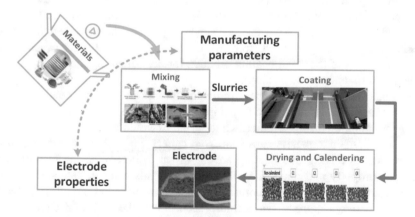

Fig. 3.4 Key steps for battery electrode manufacturing

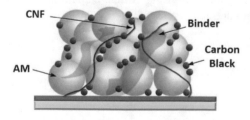

Fig. 3.5 Key material components for battery electrode manufacturing

carbon black and carbon nanofiber (CNF) are needed due to the intrinsic electronic conductivity of solo active material which is usually insufficient. In addition, polymeric bind also plays a vital role in providing mechanical cohesion [2]. In reality, three types of binders including the polyvinylidene-fluoride (PVDF), polyethylene co-ethyl-acrylate co-maleic-anhydride (TPE) and hydrogenated-nitrile-butadiene-rubber (HNBR) are generally adopted owing to their exceptional chemical stability and reliable binding properties.

After preparing proper materials, these materials will be mixed within the soft blender to produce the homogenous slurry during the mixing step. It should be known that the slurry rheological property such as active material mass content, solid-to-liquid ratio and viscosity play pivotal roles in the further coating and drying stages [23]. During coating, the slurry will be coated on the current collector made of mental foil by a coating machine. In general, copper foil is used for anode while aluminium foil is utilized for cathode. Here the coating speed is usually set as a constant value while the comma gap of coating machine would be adjusted to generate shear force that significantly affects the thickness of coating products. Defects such as the pinhole, agglomerate, and non-uniformity may happen during the mixing process and coating process, which would further decrease the electrochemical performance of the final battery such as its maximum capacity [24]. Then the wet coating product will be dried within an oven with the preset drying temperature and speed, which would highly affect the electrode's mechanical as well as electrochemical properties [25]. After that, a calendaring process is performed through the mechanical pressure generated from two cylindrical rolls to further evaporate the residual solvent of the dried coating product, further benefiting the energy density, adhesion of manufactured electrodes.

It should be noted that all these individual processes (mixing, coating, drying, calendaring) generally need specific equipment such as the mixer, coating, and dryer. Numerous manufacturing parameters and variables would be generated during these processes. Some parameters particular from battery electrode early manufacturing stages such as mixing and coating are important to determine the manufactured electrode property and must be well analysed. In this context, data science technology could first benefit the exploration of these parameters and variables involved in producing battery electrode. Furthermore, according to the well-trained data science model, the properties of manufactured electrode can be predicted without making the cells. This could be an effective and intelligent solution to guide the smart management of battery manufacturing, further benefitting the battery manufacturers to optimize their production process.

3.3.2 Case 1: Battery Electrode Mass Loading Prediction with GPR

To illustrate how to design a data science framework to benefit battery electrode manufacturing, a data science case study through deriving Gaussian process regression (GPR)-based framework to predict battery electrode mass loading is given [26]. Through incorporating automatic relevance determination (ARD) kernel, this GPR framework is able to directly quantify the importance of four intermediate manufacturing parameters and analyse their influences on the prediction of battery electrode mass loading.

Deriving from the Bayesian theory, GPR can be seen as a random process to undertake the nonparametric regression with the Gaussian processes [27]. That is, for any inputs, the corresponding probability distribution over function $f(x)$ follows the Gaussian distribution as:

$$f(x) \sim \text{GPR}\big(m(x), k\big(x, x'\big)\big) \tag{3.1}$$

where $m(x)$ and $k(x, x')$ denote the mean and covariance functions respectively, and expressed by:

$$\begin{cases} m(x) = E(f(x)) \\ k\big(x, x'\big) = E\big[\big(m(x) - f\big(x'\big)\big)\big(m(x) - f\big(x'\big)\big)\big] \end{cases} \tag{3.2}$$

Here $E()$ represents the expectation value. It is worth noting that in practice, $m(x)$ is generally set to be zero for simplifying calculation process [27]. $k(x, x')$ is also named as the kernel function to explain the relevance degree between a target observation of the training data set and the predicted output based on the similarity of the respective inputs.

In a regression issue, the prior distribution of outputs y can be expressed by:

$$y \sim N\big(0, k\big(x, x'\big) + \sigma_n^2 I_n\big) \tag{3.3}$$

$N()$ indicates a normal distribution. σ_n is the noise term. Supposing there exists a same Gaussian distribution between the testing set x' and training set x, the predicted outputs y' would follow a joint prior distribution with the training output y as:

$$\begin{bmatrix} y \\ y' \end{bmatrix} \sim N\left(0, \begin{bmatrix} k(x, x) + \sigma_n^2 I_n & k\big(x, x'\big) \\ k\big(x, x'\big)^T & k\big(x', x'\big) \end{bmatrix}\right) \tag{3.4}$$

where $k(x, x), k(x', x')$, and $k(x, x')$ represent the covariance matrices among inputs from training set, testing set, as well as training and testing sets, respectively.

In order to guarantee the performance of GPR, some hyperparameters θ existing in the covariance function require to be optimized by the n points in the training process. One efficient optimization solution is to minimize the negative log marginal likelihood $L(\theta)$ [28] as:

$$\begin{cases} L(\theta) = \frac{1}{2}\log[\det \lambda(\theta)] + \frac{1}{2}y^T\lambda^{-1}(\theta)y + \frac{n}{2}\log(2\pi) \\ \lambda(\theta) = k(\theta) + \sigma_n^2 I_n \end{cases} \tag{3.5}$$

After optimizing the hyperparameters of GPR, the predicted output y' can be obtained at dataset x' through calculating the corresponding conditional distribution $p(y'|x', x, y)$ as:

$$p(y'|x', x, y) \sim N(y'|\overline{y}', \text{cov}(y')) \tag{3.6}$$

with

$$\begin{cases} \overline{y}' = k(x, x')^T [k(x, x) + \sigma_n^2 I_n]^{-1} y \\ \text{cov}(y') = k(x', x') - k(x, x')^T [k(x, x) + \sigma_n^2 I_n]^{-1} k(x, x') \end{cases}$$

where $\overline{y}'$ stands for the corresponding mean values of prediction. $\text{cov}(y')$ denotes a variance matrix to reflect the uncertainty range of these predictions. More details on these equations of GPR can be found in [27].

3.3.2.1 ARD Kernel for Feature Selection

As two early stages of battery electrode manufacturing chain, mixing and coating could play pivotal roles in determining the mass loading of battery electrode, further affecting the performance of final manufactured cell. In this case study, a GPR-based data science model is derived to well predict battery electrode mass loading and quantify the importance weights of four battery manufacturing parameters of interest, as illustrated in Fig. 3.6. Specifically, these four parameters include three mixing feature variables: mass content (MC), solid-to-liquid ratio (STLR), slurry viscosity (V), and one coating process parameter: comma gap (CG). GPR is integrated with four ARD structure-based kernel functions, which are derived as follows:

$$k_{\text{ARDEX}}(i, i') = \sigma_{\text{EX}}^2 \exp\left[-\left(\frac{\|i_{\text{MC}} - i'_{\text{MC}}\|}{\sigma_{\text{MC}}} + \frac{\|i_{\text{STLR}} - i'_{\text{STLR}}\|}{\sigma_{\text{STLR}}} + \frac{\|i_{\text{CG}} - i'_{\text{CG}}\|}{\sigma_{\text{CG}}} + \frac{\|i_V - i'_V\|}{\sigma_V}\right)\right] \tag{3.7}$$

$$k_{\text{ARDSE}}(i, i') = \sigma_{\text{SE}}^2 \exp\left[-\frac{1}{2}\left(\frac{\|i_{\text{MC}} - i'_{\text{MC}}\|^2}{\sigma_{\text{MC}}^2} + \frac{\|i_{\text{STLR}} - i'_{\text{STLR}}\|^2}{\sigma_{\text{STLR}}^2} + \frac{\|i_{\text{CG}} - i'_{\text{CG}}\|^2}{\sigma_{\text{CG}}^2} + \frac{\|i_V - i'_V\|^2}{\sigma_V^2}\right)\right] \tag{3.8}$$

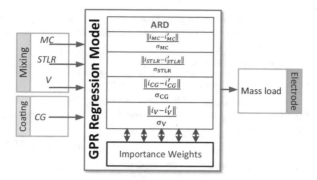

Fig. 3.6 GPR-based machine learning model to predict electrode mass load and analyse feature importance weights, reprinted from [26], with permission from Elsevier

$$\begin{cases} k_{ARDM3/2}(i, i') = \sigma_{M3/2}^2 \left(1 + \sqrt{3}r\right) \exp\left(-\sqrt{3}r\right) \\ r = \sqrt{\dfrac{\|i_{MC}-i'_{MC}\|^2}{\sigma_{MC}^2} + \dfrac{\|i_{STLR}-i'_{STLR}\|^2}{\sigma_{STLR}^2} + \dfrac{\|i_{CG}-i'_{CG}\|^2}{\sigma_{CG}^2} + \dfrac{\|i_V-i'_V\|^2}{\sigma_V^2}} \end{cases} \tag{3.9}$$

$$\begin{cases} k_{ARDM5/2}(i, i') = \sigma_{M5/2}^2 \left(1 + \sqrt{5}r + \frac{5}{3}r^2\right) \exp\left(-\sqrt{5}r\right) \\ r = \sqrt{\dfrac{\|i_{MC}-i'_{MC}\|^2}{\sigma_{MC}^2} + \dfrac{\|i_{STLR}-i'_{STLR}\|^2}{\sigma_{STLR}^2} + \dfrac{\|i_{CG}-i'_{CG}\|^2}{\sigma_{CG}^2} + \dfrac{\|i_V-i'_V\|^2}{\sigma_V^2}} \end{cases} \tag{3.10}$$

where all these ARD structure-based kernels have the same series of input items as $i = \{i_{MC}, i_{STLR}, i_{CG}, i_V\}$. $\sigma_{MC}, \sigma_{STLR}, \sigma_{CG},$ and σ_V are four hyperparameters to reflect the relevancies and importance of manufacturing parameters including MC, STLR, CG, and viscosity, respectively. According to these defined ARD kernels, the workflow of designing ARD kernel-based GPR model for feature selection and prediction of manufactured battery electrode mass loading is detailed in Table 3.2.

It should be known that taking the exponential of the negative learned θ_{hp} and normalizing W is a common solution to make the quantified importance weights of manufacturing parameters become more convenient for comparison. Other weighting strategies such as the inverse of θ_{hp} could lead the final quantified weight values slightly change, but it would not influence the importance ranking trend of these parameters. The manufacturing parameter item with higher ranking (larger normalized weight value) means this parameter is more crucial than others in predicting battery electrode mass loading. In light of this, all four parameter weights are uniformly quantified in this case study. Following the workflow, the weights of input items including the MC, STLR, viscosity from mixing and the CG from coating could be directly quantified to reflect their importance. Then the reliable feature selections can be carried out based upon these quantified importance weights. Moreover, the designed GPR model with ARD kernel structure is able to generate an explanatory subset of features by setting different hyperparameters for all inputted variable

Table 3.2 Detailed workflow of designing ARD kernel-based GPR for feature selection and prediction of manufactured battery electrode mass loading, reprinted from [26], with permission from Elsevier

1: Procedure: data preprocess and GPR model training
a. Remove outliers of raw electrode manufacturing data
b. Prepare predefined inputs–output matrix for GPR training. Here the inputs are four interested parameters from mixing and coating while output is the related mass loading of manufactured battery electrode
c. Optimize hyperparameters θ_{hp} of ARD kernel within GPR to minimize its $L(\theta_{hp})$ through fourfold cross-validation. Then the well-trained GPR model $\text{GPR}_{ARD}(\theta_{hp})$ can be obtained
2: **End procedure**
3: **Procedure:** feature selections and battery electrode mass loading prediction
a. Based on the optimized θ_{hp}, the weights of manufacturing parameters are calculated by: $$W = \exp(-\theta_{hp})$$ where $\theta_{hp} = \left\{ \sigma_{MC}, \sigma_{STLR}, \sigma_{CG}, \sigma_V \right\}$ stand for the well-optimized hyperparameters of electrode manufacturing parameters MC, STLR, CG, and viscosity, respectively. $W = \left\{ W_{MC}, W_{STLR}, W_{CG}, W_V \right\}$ are their weights by taking the exponential of negative hyperparameters b. Normalize these weights to quantify feature importance as: $$W = W/(W_{MC} + W_{StLr} + W_{CG} + W_V)$$ c. For a new observation i' from testing dataset, its related electrode mass loading y' can be predicted by using the well-established $\text{GPR}_{ARD}(\theta_{hp})$ as: $$y' = \text{GPR}_{ARD}(i' | \theta_{hp})$$
4: **End procedure**

items, further benefitting the improvement of performance and generalization ability of battery electrode mass loading prediction.

3.3.2.2   Results and Discussions

To well analyse and evaluate the feature selection and regression performance of GPR with various ARD kernels, two tests including one with all four manufacturing parameter items and another with reduced manufacturing parameter items are carried out.

First, to quantify the importance weights of interested manufacturing parameters on the prediction of manufactured battery electrode mass loading, all parameter items including MC, STLR, CG, and viscosity are adopted as the inputs for GPR models. The prediction performance of all four ARD kernels is evaluated using fourfold cross-validation. Figure 3.7 and Table 3.3 show these prediction results and the related performance indicators for these four GPRs with different ARD kernels, respectively.

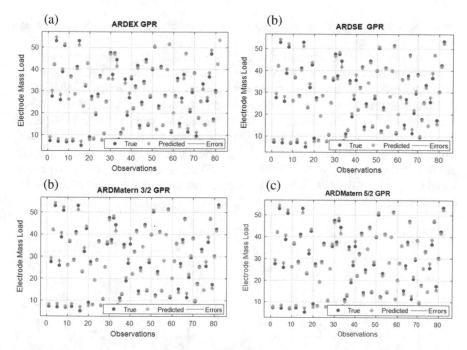

Fig. 3.7 Manufactured battery electrode mass loading prediction results by using all four items: **a** ARDEX kernel, **b** ARDSE kernel, **c** ARDMatern3/2 kernel, **d** ARDMatern5/2 kernel, reprinted from [26], with permission from Elsevier

Table 3.3 Performance indicators for electrode mass loading prediction by using all items, reprinted from [26], with permission from Elsevier

Kernel types	ARDEX	ARDSE	ARDMatern3/2	ARDMatern5/2
Training time (s)	13.547	13.687	14.154	14.487
MAE (mg/cm^2)	0.946	0.913	0.911	0.875
RMSE (mg/cm^2)	1.177	1.126	1.124	1.084

Obviously, most observations in Fig. 3.7 well match the outputs from four GPR models, indicating that all these four ARD kernel functions could provide satisfactory prediction results for the manufactured battery electrode mass loading. Quantitatively, due to the simplest structure, the training phase of GPR with ARDEX kernel can be finished within 13.547 s, while its prediction accuracy is the worst with the RMSE of 1.177 mg/cm^2. In the contrary, GPR with the ARDMatern5/2 kernel leads to the longest training time of 14.487 s (6.9% increase), but its RMSE becomes the lowest one with 1.084 mg/cm^2 (7.9% decrease). However, the training time of all these kernel functions is within 14.5 s, implying that the computational efforts of these four ARD kernels are all acceptable. In conclusion, after adopting all four manufacturing parameters as the input items for the ARD kernel-based GPR model, expected accuracy could be achieved for the manufactured battery electrode mass loading predictions.

To further reflect the deviations of electrode mass loading prediction results, the predicted versus actual plots (PVAPs) for GPRs with all four ARD kernel functions are shown in Fig. 3.8. In theory, for the observations on the left or right of PVAPs, the furthest from the average value could produce the most leverages and effectively pull the prediction line towards that observation. For a model with good prediction performance, the observations should get close to the perfect prediction line. According to Fig. 3.8, all observations can be clustered around the perfect prediction lines without obvious outliers, implying that GPRs with ARD kernel functions are able to provide satisfactory electrode mass loading prediction results with a few deviations for the case of using all manufacturing parameter items.

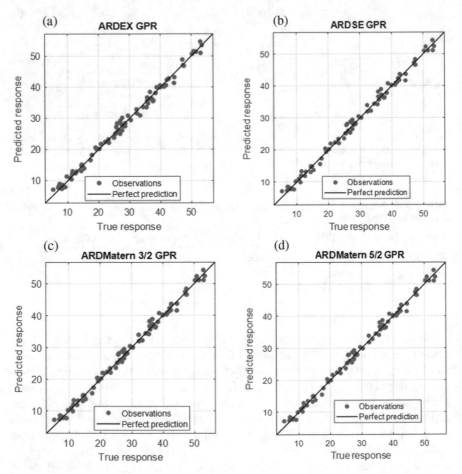

Fig. 3.8 Predicted versus actual plots by using all four items: **a** ARDEX kernel, **b** ARDSE kernel, **c** ARDMatern3/2 kernel, **d** ARDMatern5/2 kernel, reprinted from [26], with permission from Elsevier

Next, based upon the well-optimized GPR models, the normalized importance weights from hyperparameters of all four ARD kernel functions are plotted in Fig. 3.9. Interestingly, although there exist differences among each feature item, the trend of feature importance weights is similar for all GPRs with four ARD kernel functions. Specifically, CG always presents the largest important weight, which is nearly twice as large as that of STLR. In the contrary, viscosity gives the smallest importance weight (nearly five times less than the CG). The STLR's importance weights are slightly larger than MC's for all four ARD kernel functions. This finding signifies that among these four electrode manufacturing parameters, CG is the most important item and must be selected for battery electrode mass loading prediction. STLR and MC are the second and third important parameter items. In contrast, viscosity makes the least contribution to predicting the mass loading of the manufactured battery electrode.

To further reflect the electrode mass loading prediction performance of ARD kernel functions, four GPRs with the related conventional kernel functions of EX, SE, Matern3/2 and Matern5/2 are utilized as the benchmarks for comparison purposes. It should be known that these convention kernels without ARD structure cannot directly quantify the importance weights of manufacturing parameters of interest as all four parameter items' hyperparameters within kernel are same. Figure 3.10 and Table 3.4 illustrate the related electrode mass loading prediction results and corresponding

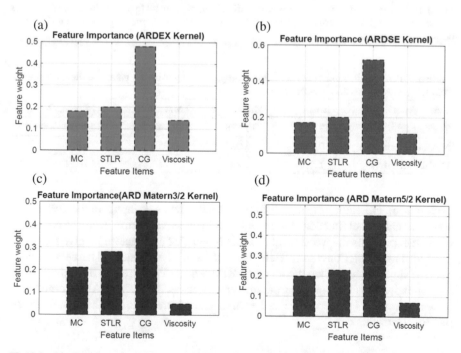

Fig. 3.9 Obtained importance weights by using different ARD kernels: **a** ARDEX kernel, **b** ARDSE kernel, **c** ARDMatern3/2 kernel, **d** ARDMatern5/2 kernel, reprinted from [26], with permission from Elsevier

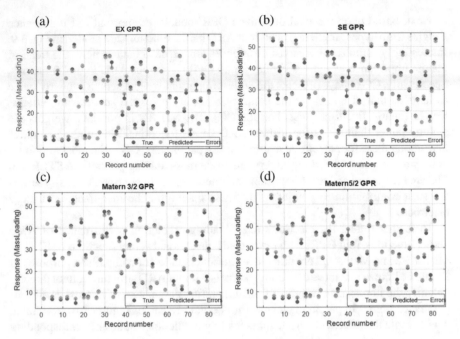

Fig. 3.10 Comparison results by using conventional kernels with all four manufacturing parameter items: **a** EX kernel, **b** SE kernel, **c** Matern3/2 kernel, **d** Matern5/2 kernel, reprinted from [26], with permission from Elsevier

Table 3.4 Performance indicators for kernel comparisons by using all parameter items, reprinted from [26], with permission from Elsevier

Kernel types	EX	SE	Matern3/2	Matern5/2
Training time (s)	13.323	13.586	13.964	13.959
MAE (mg/cm^2)	0.994	0.983	0.958	0.907
RMSE (mg/cm^2)	1.317	1.292	1.184	1.151

performance indicator of using all four parameter items as inputs to the GPRs with conventional kernel functions based on fourfold cross-validation. It can be noted that EX kernel presents the worst results, while Matern3/2 and Matern5/2 give the better mass loading prediction results. Here the RMSEs of all conventional kernel functions become slightly worse than the related ARD kernels (5% increase). Therefore, apart from the advantage of explaining parameter importance, GPRs with ARD kernels also present competent performance in the mass loading prediction of the manufactured battery electrode.

3.3.3 Case 2: Battery Electrode Property Classification with RF

In this study, another data science-based solution through deriving an improved random forest (RF) classification model is given to effectively classify electrode properties and quantify the feature importance and correlations among four early manufacturing parameters, as shown in Fig. 3.11 [29]. To be specific, the model inputs include slurry active material mass content (AMMC), solid-to-liquid ratio (StoLR) and viscosity from the mixing stage, and one process parameter of coating named comma gap (CG). The model output is the labelled classes of electrode mass loading or porosity.

3.3.3.1 RF Technique and Feature Analyses Solutions

As a typical ensemble learning solution, numerous individual decision trees (DTs) are integrated within RF, as illustrated in Fig. 3.12. In general, classification and regression tree (CART) is adopted as a DT of RF owing to its simplification and nonparametric behaviours. The main process for RF training is to produce various de-correlated DTs. To reduce its variance, an overlap sampling method called "bagging" is utilized here [30]. Additionally, to restrain correlations of these DTs, the best split of each node would be got by randomly choosing m subset features from all M features. In this context, DTs could be built without pruning, resulting in a relatively low computational effort. Besides, after adopting various bootstrap samples and node features, RF's noise immunity is improved by averaging different DTs.

To effectively quantify the importance of battery manufacturing parameters of interest, two different types of feature importance (FI) including the unbiased FI and gain improvement FI are utilized in this study. Table 3.5 illustrates the detailed process to obtain the unbiased FI.

Apart from the unbiased FI, another effective FI is obtained by summing the gain improvements of Gini impurity changes. For a classification, Gini impurity is used

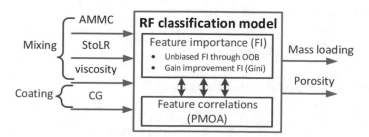

Fig. 3.11 RF-based classification model structure for classifying and analysing manufactured battery electrode properties, reprinted from [29], with permission from IEEE

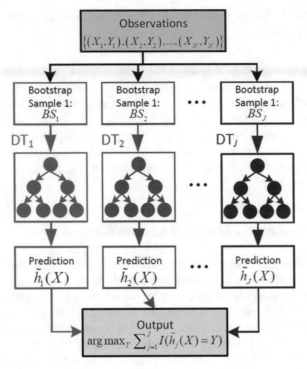

Fig. 3.12 Schematic of RF model, reprinted from [29], with permission from IEEE

Table 3.5 Detailed process to obtain the unbiased FI

1: **Procedure:** to obtain the unbiased importance of features $x_k (k = 1 : M)$
2: (Obtain $\tilde{Y}_{i,j}$) For $i = 1 : N$
a. Supposing BS_j represents the jth bootstrap sample, $A_i = \{j : (X_i, Y_i) \notin BS_j\}$, and J_i means the cardinality of A_i
b. Obtaining $\tilde{Y}_{i,j} = \tilde{h}_j(X_i)$ for all $j \in A_i$
3: (Obtain $\widetilde{Y'}_{i,j}$) For $j = 1 : J$
a. Supposing $B_j = \{i : (X_i, Y_i) \notin BS_j\}$
b. Randomly permuting x_k from data samples $\{X_i : i \in B_j\}$ to generate $C_j = \{X'_i : i \in B_j\}$
c. Obtaining $\widetilde{Y'}_{i,j} = \tilde{h}_j(X'_i)$ for all $i \in B_j$
4: For $i = 1 : N$
Calculating the local FI $\text{LFM}_i(x_k)$ of x_k as: $$\text{LFM}_i(x_k) = \frac{1}{J_i} \sum_{j \in A_i} I(Y_i \neq \widetilde{Y'}_{i,j}) - \frac{1}{J_i} \sum_{j \in A_i} I(Y_i \neq \tilde{Y}_{i,j})$$
5: Obtaining the overall unbiased importance $\text{OFM}(x_k)$ of feature x_k as: $$\text{OFM}(x_k) = \frac{1}{N} \sum_{i=1}^{i=N} \text{LFM}_i(x_k)$$
End procedure

Table 3.6 Detailed process to obtain Gain improvement FI

1: **Procedure**: to obtain the Gain improvement FI of features $x_k (k = 1 : M)$
2: (Obtain $\Delta\text{Gini}(\tau, x_k)$) For $j = 1 : J$
a. For a node τ of DT_j, calculating its Gini impurity $\text{Gini}(\tau)$ as: $$\text{Gini}(\tau) = 1 - \sum_{d=1}^{D} p_k^2$$ where D represents the number of classes, $p_k = n_k/n$ means the fraction of n_k samples out of total n samples
b. Calculating all Gini impurities $\text{Gini}(\tau, x_i)$ under the case of chosen feature x_i by: $$\text{Gini}(\tau, X) = \frac{|\tau_l|}{|\tau|}\text{Gini}(\tau_l) + \frac{|\tau_r|}{|\tau|}\text{Gini}(\tau_r)$$ where τ_l and τ_r represent the left and right children of the current node τ, respectively; $|\tau|$, $|\tau_l|$ and $|\tau_r|$ are the numbers of records in τ, τ_l and τ_r, respectively
c. Calculating the Gini reduce $\Delta\text{Gini}(\tau, X)$ of all selected X by: $\Delta\text{Gini}(\tau, X) = \text{Gini}(\tau) - \text{Gini}(\tau, X)$
d. Comparing $\Delta\text{Gini}(\tau, X)$ to get the optimal split feature x_k at the specific node τ. Recording its Gini reduce $\Delta\text{Gini}(\tau, X_k)$
3: (Obtain $I_G(x_k)$) For $j = 1 : J$
a. Accumulating the recorded $\Delta\text{Gini}(\tau, x_k)$ for all utilized nodes (ANs) in all trees (ATs) by: $$S\Delta\text{Gini}(x_k) = \sum_{\text{ATs}} \sum_{\text{ANs}} \Delta\text{Gini}(\tau, x_k)$$ where $S\Delta\text{Gini}(x_k)$ represent the summed gain improvement according to x_k's Gini variations
b. Calculating the overall gain improvement $I_G(x_k)$ of parameter x_k as: $$I_G(x_k) = \frac{1}{N_k} S\Delta\text{Gini}(x_k)$$ where N_k stands for the cardinality of $S\Delta\text{Gini}(x_k)$
4: **End procedure**

to reflect how well a potential split is in a specific node of DT [31]. The detailed process to calculate the gain improvement FI is described in Table 3.6.

In addition, quantifying the correlations of different manufacturing parameter pairs is also vital to understanding battery manufacturing. To achieve this, the predictive measure of association (PMOA) is designed. The basic idea of calculating PMOA is through the comparison of all potential splits with the optimal one which is observed during DT training. Then the best surrogate decision split could generate the maximum PMOA value to reflect their corresponding correlations. In this context, PMOA is able to reflect the similarity between various decision rules to split observations. Supposing x_e and x_g are two interested feature variables ($e \neq g$), the PMOA between the optimal split $x_e < u$ and surrogate split $x_g < v$ can be calculated as:

$$\text{PMOA}_{e,g} = \frac{\min(\text{Pl}, \text{Pr}) - 1 + \text{Pl}_e l_g + \text{Pr}_e r_g}{\min(\text{Pl}, \text{Pr})} \tag{3.11}$$

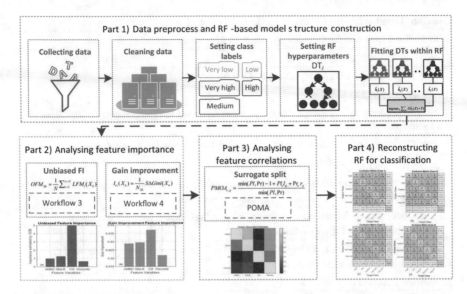

Fig. 3.13 Data science framework of designing RF-based model for the classification and sensitivity analysis of manufacturing parameters, reprinted from [29], with permission from IEEE

where l and r denote the left and right children of node; Pl is the observation proportion of $x_e < u$ while Pr means the observation proportion of $x_e \geq u$; $\text{Pl}_e l_g$ represents the observation proportion of $x_e < u$ and $x_g < v$; $\text{Pr}_e r_g$ is the observation proportion of $x_e \geq u$ and $x_g \geq v$. PMOA should be less than 1, larger PMOA implies the related feature pair has higher correlations.

Figure 3.13 illustrates the data science framework of designing the RF-based model for the classification and sensitivity analysis of specific battery electrode manufacturing parameters. This framework can be divided into four main parts as follows:

Part 1. Data preprocess and RF-based model construction: After collecting relevant data from battery electrode manufacturing chain, the outliers are firstly removed and the class labels are set for outputs. In this study, both mass loading and porosity of manufactured battery electrode are set with five class labels. Then the inputs–output pairs as illustrated in Fig. 3.11 are generated to train all DTs within RF. In theory, RF classification model has two main hyperparameters, the amount of DTs (J) and the number of feature items during each split (m), need to be tuned. Some points should be considered for the tunning of these two hyperparameters: On the one hand, in theory, higher accuracy and generalization ability can be obtained when increasing J. However, too many DTs could also increase the computational burden of derived RF. On the other hand, both the performance and correlations of DTs within RF model could be affected by m. High value of m could benefit DT's strength but also lead these DTs become more correlated. To obtain suitable hyperparameters in this study, an effective tunning approach called randomized search [32] is utilized.

Part 2. Feature importance analysis: in this part, to quantify the feature importance of all manufacturing parameters of interest and analyse their effects on the classification of electrode mass loading and porosity, two quantitative metrics including the unbiased FI and the gain improvement FI are adopted. Specifically, the unbiased FI is calculated based on the process in Table 3.5, while the gain improvement FI is obtained based on the process in Table 3.6.

Part 3. Feature correlations analysis: After quantifying the importance of mixing and coating parameters, the PMOA values of each parameter pair are calculated by Eq. (3.11) and visualized as a $M \times M$ heat map. Then the correlations between each two manufacturing parameters can be quantified by these PMOAs. Larger PMOAs theoretically indicate that higher correlations exist between parameter pairs. The PMOAs of two manufacturing parameters could be different in a heat map, relying on which manufacturing parameter causes DTs' optimal spit firstly.

Part 4. RF classification model reconstruction: After the comparison of FI and the analysis of parameter correlations, the most important manufacturing parameters that affect battery electrode property classification results are selected. Then the RF could be reconstructed with reduced parameters for new electrode property classifications.

Following these steps, an effective RF model-based data science framework can be designed to not only analyse the importance and correlations of early manufacturing parameters from mixing and coating, but also well classify the mass loading and porosity of manufactured battery electrode into proper categories.

3.3.3.2 Electrode Mass Loading Analysis

Feature analyses: for the electrode mass loading classification, following the steps from Tables 3.5 and 3.6, both unbiased FI and gain improvement FI of all four feature variables could be quantified, as shown in Fig. 3.14. Interestingly, although the value levels between these two FIs are significantly different, a similar trend for all feature variables can be obtained. Obviously, CG presents much higher values for both unbiased FI with 4.78 and gain improvement FI with 0.037, indicating that this feature variable is the most important one. StoLR and AMMC provide the second and third larger values for both two types of FIs. The viscosity gives the smallest values of unbiased FI with 0.67 and gain improvement FI with 0.022, indicating that this feature contributes the least to classify manufactured electrode mass loading.

Next, the heat map to reflect the PMOAs of all feature pairs are created for evaluating the correlations among four features for electrode mass loading case, as illustrated in Fig. 3.15. Quantitatively, AMMC and StoLR achieve the largest correlations with the PMOA of 0.72. This correlation output is very useful as the obtained results are consistent with the observations from manufacturing experiments, but this study demonstrates how an RF-based data science framework can support the interpretation of correlations among interested manufacturing feature variables. This

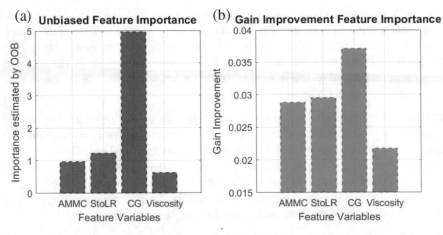

Fig. 3.14 FI for battery mass loading. **a** Unbiased FI based on OOB **b** FI based on gain improvement, reprinted from [29], with permission from IEEE

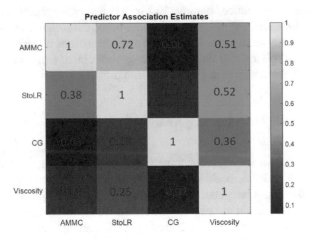

Fig. 3.15 Heat map to reflect feature correlations for battery mass loading case, reprinted from [29], with permission from IEEE

could benefit engineers to efficiently understand and manage their battery electrode manufacturing chain.

Electrode mass loading classification: To evaluate battery electrode mass loading classification results, a test by using the derived RF model with all features as inputs is carried out first. According to its CM in Fig. 3.16, a satisfactory classification accuracy rate with 90.2% is achieved. Quantitatively, the best classification results are classes "very high" and "very low" with 100% P_{rate}, while the worst classification result is the "low" class with 72.7% P_{rate}. This is mainly due to two observations

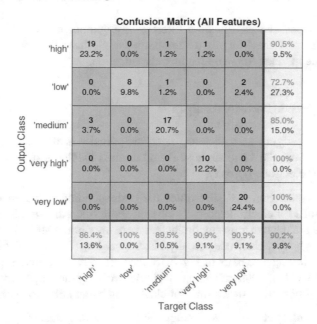

Fig. 3.16 Confusion matrix for mass load classification results by using all features, reprinted from [29], with permission from IEEE

being incorrectly classified as "very low" and one observation being classified as "medium" in such a case.

Next, to further evaluate the effects of each manufacturing feature on the classification of manufactured electrode mass loading, four different cases with various combinations of three feature items are tested. To be specific, Case 1 contains CG, AMMC and StoLR. Case 2 includes CG, AMMC and viscosity. Case 3 includes CG, StoLR, and viscosity. Case 4 is composed of AMMC, StoLR, and viscosity. The performance metrics of all these four cases are illustrated in Table 3.7. It can be seen that Case 1 presents the best classification result with 86.6% $macroP$, 89.8% $macroR$ and 90.0% $macroF1$, which are just 3.3, 1.9 and 0.1% less than those from the case of using all feature items. This implies that using CG, AMMC, and StoLR is

Table 3.7 Performance indicators for battery electrode mass loading classification, reprinted from [29], with permission from IEEE

Cases	$macroP$ (%)	$macroR$ (%)	$macroF1$ (%)
All features	89.6	91.5	90.1
Case 1	86.6	89.8	90.0
Case 2	84.6	84.8	84.6
Case 3	83.9	85.2	84.3
Case 4	35.3	36.2	35.4

sufficient for mass load classification. Interestingly, without involving CG, the performance metrics of Case 4 largely decrease, indicating that CG plays a significantly important role in electrode mass loading classification.

3.3.3.3   Electrode Porosity Analysis

The test of classifying manufactured electrode porosity is also carried out. In this test, the inputs are the same as those from the mass loading test, while the output here becomes electrode porosity.

Feature analyses: Fig. 3.17 illustrates the corresponding unbiased FI and gain improvement FI for electrode porosity classification. According to Fig. 3.17, StoLR and viscosity are the two most important feature items while AMMC is the worst one for classifying battery electrode porosity. Next, according to the association estimates of corresponding feature pairs in Fig. 3.18, the pair of AMMC and StoLR presents the highest PMOA of 0.84, indicating that these two early manufacturing parameters present strong potential correlations for classifying battery electrode porosity.

Electrode porosity classification: After using all four manufacturing parameters as inputs, its CM to reflect the classification results of manufactured battery electrode porosity is illustrated in Fig. 3.19. This test could provide a classification accuracy of 70.7%, which is mainly caused by some mismatched observations particular for the class label "high". In comparison with the electrode mass loading case, it can be seen that the quality of electrode porosity classification cannot be fully determined by these four manufacturing parameters of interest.

Next, to further investigate the effects of these manufacturing parameters on electrode porosity classification, four tests with similar parameter combination cases

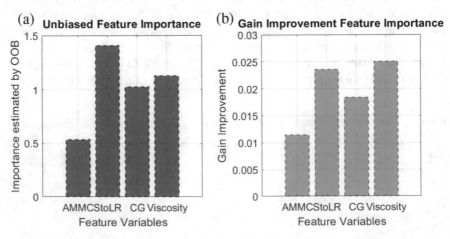

Fig. 3.17 Feature importance for battery porosity. **a** Unbiased FI based on OOB **b** FI-based on gain improvement, reprinted from [29], with permission from IEEE

Predictor Association Estimates
AMMC
StoLR
CG
Viscosity
1
0.84
0.28
0.53
0.26
1
0.46
1
0.55
0.26
0.31
1
AMMC
StoLR
CG
Viscosity
1
0.9
0.8
0.7
0.6
0.5
0.4
0.3
0.2
0.1

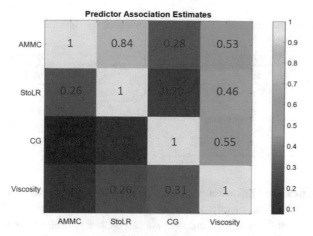

Fig. 3.18 Heat map to reflect feature correlations for battery porosity case, reprinted from [29], with permission from IEEE

Confusion Matrix (All Features)
Output Class
'high'
'low'
'medium'
'very high'
'very low'
2 2.4% | 1 1.2% | 3 3.7% | 1 1.2% | 0 0.0% | 28.6% 71.4%
0 0.0% | 30 36.6% | 3 3.7% | 0 0.0% | 4 4.9% | 81.1% 18.9%
2 2.4% | 4 4.9% | 9 11.0% | 0 0.0% | 0 0.0% | 60.0% 40.0%
1 1.2% | 1 1.2% | 0 0.0% | 3 3.7% | 0 0.0% | 60.0% 40.0%
0 0.0% | 4 4.9% | 0 0.0% | 0 0.0% | 14 17.1% | 77.8% 22.2%
40.0% 60.0% | 75.0% 25.0% | 60.0% 40.0% | 75.0% 25.0% | 77.8% 22.2% | 70.7% 29.3%
'high'
'low'
'medium'
'very high'
'very low'
Target Class

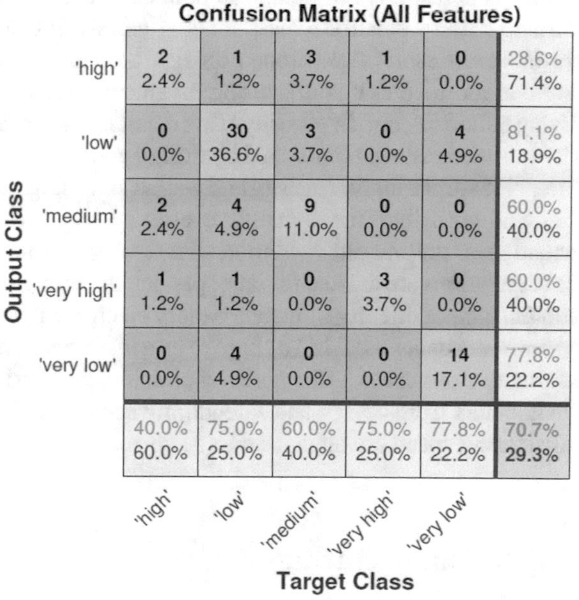

Fig. 3.19 Confusion matrix for porosity classification results by using all features, reprinted from [29], with permission from IEEE

as those from mass loading are compared here, while their performance metrics are illustrated in Table 3.8. Specifically, through adopting the three most important manufacturing parameters (StoLR, CG, and viscosity), Case 3 shows the best classification result with 59.4% $macroP$, 60.8% $macroR$, and 59.7% $macroF1$. However, the

Table 3.8 Performance indicators for battery electrode porosity classification, reprinted from [29], with permission from IEEE

Cases	macroP (%)	macroR (%)	macroF1 (%)
All features	61.5	65.6	66.4
Case 1	54.9	54.7	54.9
Case 2	53.8	48.3	50.6
Case 3	59.4	60.8	59.7
Case 4	67.2	56.8	54.6

overall classification performance of electrode porosity is worse than the electrode mass loading case. This fact signifies that for electrode porosity classification, more other related manufacturing parameters should be considered to further enhance its classification performance.

In summary, the electrode mass loading can be well classified based on these four manufacturing parameters with 90.1% $macroF1$ while CG plays a most pivotal role in this classification. This result is reasonable as both coating weight and thickness that would determine battery electrode mass loading are significantly affected by CG. For the classification result of electrode porosity, the $macroF1$ becomes just 66.4%, indicating that more other manufacturing parameters need to be considered for better classifying battery electrode porosity. This result is also expected as battery electrode porosity is also significantly affected by drying parameters such as temperature and pressure. Not surprisingly, AMMC and StoLR have large correlations for both mass loading and porosity. This is mainly caused by the ratio between slurry solid component and mass present direct relation with the active material property. In contrast, none direct relations are existed for other parameter pairs. In addition, active material mass content cannot directly influence battery electrode physical properties such as porosity, further causing the AMMC here to be a less important parameter. In this context, to further improve electrode porosity classification performance, more manufacturing parameters from drying and calendaring such as drying temperature, pressure, and calendaring speed are suggested to be considered.

3.4   Battery Cell Manufacturing

3.4.1   Overview of Battery Cell Manufacturing

Figure 3.20 illustrates the general steps for battery cell manufacturing, which mainly includes the processes of electrode cutting, cell assembly, electrolyte filling, forming, and testing. For battery cell manufacturing, the manufactured electrode would be first cut into different sizes for various battery types such as the coin cell, cylindrical cell, and pouch cell. There are usually two typical options including the die-cut and laser-beam cut to cut the manufactured electrode [33]. Both of them present merits and

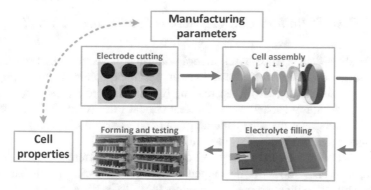

Fig. 3.20 Key steps for battery cell manufacturing

demerits. The die-cut is easier to result in the delamination of anode edge and the bending of current collector, while laser-beam cut could produce less damage to the electrodes. However, the aluminium spatter may promote the dendrite growth to further lead cell failure when laser-beam cut is utilized for cathode. For the process of cell assembly, the assembly of electrode and separator becomes a main variation source. The electrode's misalignment would highly affect both charging/discharging capacity and cell ageing behaviours [34, 35]. Specifically, the overlapping area of anode and cathode would be decreased firstly, then the dendrite will initially grow by the non-uniform Li-ions depositing at the edges [36]. Moreover, the separator here would be selected based on different characteristics. For example, its weight and thickness would affect the manufactured cell's gravimetric and volumetric performance, while its wettability and porosity would affect the electrolyte quantity and cell service life [37, 38].

After assembling the cell, the electrolyte filling and wetting would be the next steps and would also highly affect the final performance of the manufactured battery. Here the electrolyte is an ionic conductor between the active materials of electrode for ensuring ion exchange. The poorly wetted electrode should be avoided in this stage as they can result in an increased impedance and dendrite formation of the manufactured battery. Finally, as the two most time-consuming processes, the forming and testing of manufactured cells should be well designed and controlled. In theory, a stable SEI could be formed through consuming active lithium and electrolyte materials during the stage of battery formation [39]. This SEI layer could provide protections against the irreversible consumption of electrolyte and the damage to the particles of active materials [40]. To eliminate the generated gaseous product, the manufactured battery cell needs to be degassed and tested after resealing. Here the battery electrochemical test is customized for intended different applications. In principle, its current and voltage need to be accurately monitored to generate accurate information on manufactured cell capacity, cycling loss, energy, and power densities.

3.4.2 Case 1: Battery Cell Capacities Prediction with SVR

In this study, a data science-based model with support vector regression (SVR) is developed to describe how the variations in assembled battery half-cell properties including the thickness (μm), mass loading (g/m^2) and porosity (%) affect the final manufactured battery capacities at different current rates (C/20, 1C and 2C) [41]. Figure 3.21 illustrates the schematic diagram of all parts and the predefined inputs–output matrix with 115 observations in total.

To predict manufactured battery cell capacities under different C-rates, an SVR-based data science model is developed. SVR belongs to a typical data science tool for both classification and regression [42]. To achieve reasonable prediction, the best regression hyperplane would be searched during the training stage of SVR. The hyperplane is determined by an orthogonal weight vector ω that could give wider margin of separations. Supposing the training dataset is noted as $TD = (X_i, Y_i)$, $i = 1, 2, \ldots, l$, $X \in R^m$, while hyperplane is $(\omega \cdot X_i + b) = 0$. To ensure all observations can be predicted well, following constraints should be satisfied as:

$$Y_i(\omega \cdot X_i + b) \geq 1, \quad i = 1, 2, \ldots, l \tag{3.12}$$

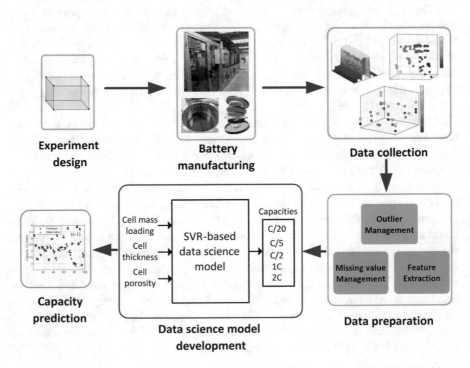

Fig. 3.21 Methodology workflow for manufacturing line modelling and predictions, reprinted from [41], with permission from Elsevier

Then the process to maximize the regression margin is defined as:

$$\begin{cases} \|\min\|\ \omega_2^2/2 \\ \text{s.t. } Y_i(\omega \cdot X_i + b) \geq 1, \quad i = 1, 2, \ldots, l \end{cases} \tag{3.13}$$

After constructing Lagrange function, this process can be expressed by the Lagrange multiplier α_i as:

$$\begin{cases} \underset{\alpha}{\min}\ \frac{1}{2}\sum_{i=1}^{N}\sum_{j=1}^{N}\alpha_i\alpha_j Y_i Y_j\left(X_i \cdot X_j\right) - \sum_{i=1}^{N}\alpha_i \\ \sum_{i=1}^{N}\alpha_i Y_i = 0 \\ \alpha_i \geq 0,\ i = 1, 2, \ldots, N \end{cases} \tag{3.14}$$

Based upon Eq. (3.14), SVR is capable of not only guaranteeing the accuracy of regression, but also maximizing the blank ranges on all sides of hyperplane [43].

In order to improve the nonlinear prediction performance of SVR, kernel functions should be coupled within SVR. Specifically, through using proper kernels, raw data from the original space could be effectively transferred to a high-dimensional space, then the SVR-based regression model could be trained by using the data from this high-dimensional space with the linear classification approach. Supposing $\phi(e)$ is a function to map the input space to a new feature space, the kernel function can be expressed by:

$$K(e, g) = \phi(e) \cdot \phi(g) \tag{3.15}$$

According to Eq. (3.14), the cost function to maximize the regression margin through involving the kernel functions becomes:

$$W(\alpha) = \frac{1}{2}\frac{1}{2}\sum_{i=1}^{N}\sum_{j=1}^{N}\alpha_i\alpha_j Y_i Y_j\left(X_i \cdot X_j\right) - \sum_{i=1}^{N}\alpha_i \tag{3.16}$$

Based upon the above discussions, kernel functions play key important roles in determining the prediction performance of SVR. It should be known that for different applications, various kernel functions would present different performances, which should be carefully selected. In this study, to well predict cell capacities of manufactured battery, an SVM-based data science model with a quadratic kernel is utilized.

After preparing the relevant battery manufacturing data, a test is carried out to predict capacities under other C-rate in between the maximum and minimum C-rates used for training purpose of SVR, even if the associated data is not used to train model. To achieve this, four items including the mass loading, thickness, porosity, and C-rate of cyclic current are utilized as the inputs of SVR, then the model will be trained based on the manufactured cell capacity data at all C-rates except a specifical one which would be used for validation purpose.

Figure 3.22 illustrates the predicted capacity values of three different cases and their relations to input items. For case (a), manufacturing data under C-rates of C/20, C/2, 1C, and 2C are utilized for training while the data with C/5 current is used for validation. The RMSE for such a case is 0.128 mAh while the corresponding R^2 is 0.98. For case (b), manufacturing data under C-rates of C/20, C/5, 1C, and 2C are used to train model while C/2 is used for validation. Here the RMSE becomes 0.128 mAh, while R^2 is 0.97. For case (c), manufacturing data under C-rates of C/20, C/2, C/5 and 2C are utilized for training while the data with 1C current is used for validation. Here the RMSE is 0.150 mAh while R^2 becomes 0.97. Based upon these results, the prediction accuracy of manufactured cell capacities under low C-rate is desirable, even if they are not included in the training data to build the SVR model. The prediction results for 1C capacity are slightly worse, which is mainly caused

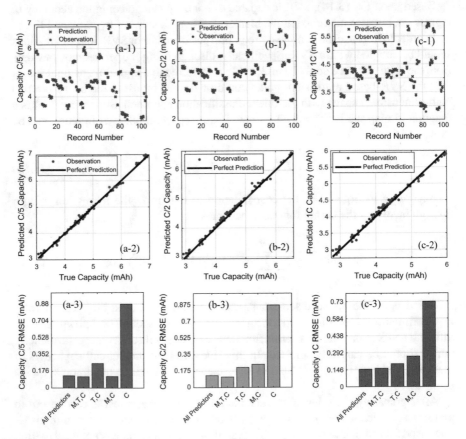

Fig. 3.22 Data science model capability for capacity interpolation, reprinted from [41], with permission from Elsevier

by its dependency on 2C capacity data presents higher variability in comparison to those with C/20, C/2, and C/5 cases.

In addition, the effects of each input item on the capacity prediction performance are also analysed. It can be seen that C-rate cannot be seen as a sole input to get satisfactory results of cell capacity prediction. Besides, the mass loading, thickness, and C-rate (M, T, C) are the input items that present the most significant effects on manufactured cell capacity prediction accuracy.

It should be known that the ability of SVR model with an extra current input to well predict manufactured cell capacity at other cyclic conditions is very useful for battery manufacturers as it could eliminate the requirements to run experiments for obtaining the corresponding data. This is more important for low C-rate cases due to the huge time-consumption of the related experiment.

3.4.3 Case 2: Battery Cell Capacity Classification with RUBoost

In the applications of battery cell manufacturing, material component formulations play a pivotal role in determining manufactured battery performance such as cell capacity. In this case study, an effective data science framework based on an advanced ensemble learning technology named RUBoost is designed to efficiently classify manufactured battery capacity considering the class imbalance issue, while the sensitivity of material component parameters of interest is also analysed [44]. The structure of this RUBoost model is shown in Fig. 3.23. To be specific, five parameters of material components including the active material weight-fraction (AMw), C65 weight-fraction (C65w), CNF weight-fraction (CNFw), Binder weight-fraction (Binderw), and Binder type are inputs of RUBoost model, while related manufactured cell capacity under C/20 is adopted as model output.

For the multiple classification applications, class imbalance is easy to occur. Two techniques including data sampling and boosting are generally adopted to alleviate the effects caused by class imbalance. For data sampling, observations from training

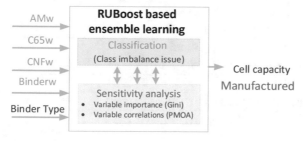

Fig. 3.23 Structure of RUBoost model for manufactured electrode properties classification, reprinted from [44], with permission from IEEE

data can be balanced by decreasing samples (undersampling) or adding samples (oversampling). In theory, as the size of training data is reduced, undersampling can benefit the computational effort but could also lead to the loss of information. On the contrary, oversampling is free of information loss. But it can bring overfitting and increase the computational burden.

On the other hand, boosting technique could be also used to overcome the issue of class imbalance. One typical boosting solution is adaptive boosting (AdaBoost) [45]. In this study, to further improve the battery cell capacity classification performance considering the class imbalance issue, a hybrid solution named random undersampling boosting (RUBoost) is derived. Supposing a training dataset $\{(X_1, C_1), (X_2, C_2), \ldots, (X_N, C_N)\}$ has N observations, X_i refers to a vector consisting of P interested items, C_i is class output with K labels, $L(X)$ means a weak learner to output a class based on X, then the detailed steps to establish the related RUBoost-based ensemble learning model is shown in Table 3.9.

Obviously, one big difference between RUBoost and AdaBoost is that a random undersampling is utilized for RUBoost in each iteration to decrease the majority class's observations to the designed percentage ($P\%$), further leading to a temporary dataset TD'_j with a new weight distribution W'_j. Then the weak learner $L'^{(j)}(X)$

Table 3.9 Detailed process to establish RUBoost data science-based model for multi-classification

1: **Procedure**: RUBoost
2: Initializing the weights of all training dataset TD as: $\qquad W_i = 1/N, \quad i = 1, 2, \ldots, N$
3: Supposing M stands for the number of all weak multiple-class learners $L'^{(j)}(X)$. For $j = 1 : M$ a. Creating a temporary training dataset TD'_j with related weights W'_j and percentage $P\%$ according to the random under sampling solution
$\qquad$ b. Fitting a $L'^{(j)}(X)$ to TD'_j according to W'_j
$\qquad$ c. Computing the $\text{err}^{(j)}$ according to TD and W_i as: $\qquad\qquad \text{err}^{(j)} = \sum_{i=1}^{N} W_i \cdot I\left(L'^{(j)}(X_i) \neq C_i\right) / \sum_{i=1}^{N} W_i$ where $\qquad I\left(L'^{(j)}(X_i) \neq C_i\right) = 1$
$\qquad$ d. Computing the weight updating factor $\alpha^{(j)}$ as: $\alpha^{(j)} = \log\left[\left(1 - \text{err}^{(j)}\right)/\text{err}^{(j)}\right] + \log(K - 1)$
$\qquad$ e. For $i = 1, 2, \ldots, N$, updating W_i as: $\qquad\qquad W_i \leftarrow W_i \cdot \exp\left[\alpha^{(j)} \cdot I\left(L'^{(j)}(X_i) \neq C_i\right)\right]$
$\qquad$ f. Renormalizing W_i
4: Outputting the final predicted class $\tilde{C}(X)$ as: $\qquad\qquad \tilde{C}(X) = \arg\max_{k} \sum_{j=1}^{M} \alpha^{(j)} \cdot I\left(L'^{(j)}(X_i) \neq C_k\right)$ where $I\left(L'^{(j)}(X_i) = k\right) = 1$
5: **End procedure**

will be well-trained by using TD'_j and W'_j. According to this undersampling way, RUBoost is able to not only provide the balanced observation for training but also results in a decreased computational effort.

Detailed procedure of using this RUBoost-based data science framework to classify cell capacity and to perform sensitivity analysis of manufacturing parameters is summarized in Fig. 3.24, which contains the following four main steps:

Step 1: Data curation and preprocess: Following a defined setting rule shown in Table 3.10, capacities of both various Li-ion battery types (LFP and LTO) are classified with three labels as low, medium, high.

Step 2: RUBoost model construction: in this step, RUBoost's hyperparameters need to be set firstly. In theory, three main hyperparameters are required for the

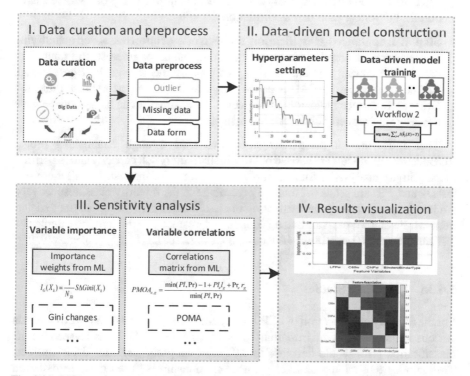

Fig. 3.24 RUBoost-based framework to conduct sensitivity analysis and classification for the battery electrode manufacturing, reprinted from [44], with permission from IEEE

Table 3.10 Class labels of manufactured battery electrode capacity

Class labels	Low	Medium	High
Capacity (LFP) [mAh/g]	13.323	13.586	13.964
Capacity (LTO) [mAh/g]	0.994	0.983	0.958

RUBoost-based data science model: the decision trees' number (M), learning rate (r), and desired percentage that is represented by minority class ($P\%$). A larger M could increase model's classification accuracy but could also result in the increased computational effort and overfitting issue. In this study, an iteration strategy based on the evaluation of classification error is performed to select proper M. Second, r means the decay speed of each learner's weight, while the class observations are balanced by $P\%$ during the training stage. As recommended by [46], setting $r = 0.1$ and $P\%$ as the minority class's percentage is a good solution.

Step 3: Sensitivity analysis: After well training the RUBoost-based model, the sensitivity analysis of variable importance and correlations could be carried out based on the Gini importance and PMOA. The detailed procedure to calculate the Gini importance refers to [31] for the readers of interest, while PMOA could be calculated based on the Eq. (3.11). Then a heat map consisting of PMOAs can be obtained to reflect each pair's correlation.

Step 4: Imbalanced classification of electrode properties: After quantifying importance and correlations of battery components of interest, the quality of battery cell capacity would be also predicted by the derived RUBoost model. The confusion matrix (CM) is adopted here as the main performance indicator. Other performance indicators from Sect. 1.3.3 in Chap. 1 can be used to evaluate the electrode property classification results.

Then following all these four steps, two tests through establishing proper RUBoost-based models are carried out to classify battery cell capacity and to perform the sensitivity analysis of four component parameters of interest. For all these two tests, r and $P\%$ are set as 0.1 and the percentage of the minority class, respectively. An iteration way through the evaluation of classification error is carried out to determine M. Fivefold cross-validation is conducted, resulting in the training sample and test sample are 110 and 28 for LFP case, 86 and 22 for LTO case, respectively.

3.4.3.1   LFP Case Studies

For the capacity classification of LFP-based battery cell, its classification error versus various M and related Gini importance are shown in Fig. 3.25. It can be seen that its classification error would decrease to 0.02 after adopting just three decision trees. This implies that an approximately linear relationship between these formulation variables and LFP-based cell capacity exists. The LFPw provides the largest Gini importance weight, while C65w presents a bit larger importance value than that of CNFw. Additionally, according to the heat map in Fig. 3.26, LFPw provides a relatively higher correlation with C65w, CNFw, and Binderw.

To evaluate the classification performance, the CMs of the cases of using all four manufacturing component parameters and three most important parameters of LFP-based battery cell capacity are generated, as illustrated in Fig. 3.27. Here the

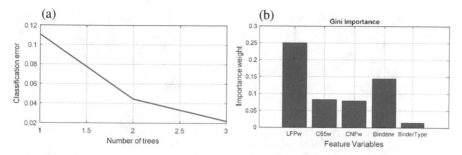

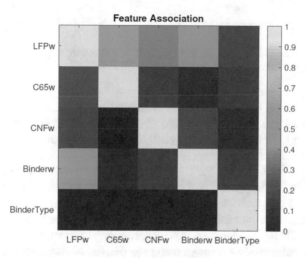

Fig. 3.25 Classification error and Gini importance for LFP capacity case. **a** classification error versus M; **b** Gini importance, reprinted from [44], with permission from IEEE

Fig. 3.26 Feature association for LFP capacity classification case, reprinted from [44], with permission from IEEE

$microF1$ of both all and reduced parameter cases reach 97.8% and 95.7%, respectively, indicating that a fantastic classification performance could be obtained through using the derived RUBoost model to classify LFP-based battery cell capacity.

3.4.3.2   LTO Case Studies

Next, the test through using the designed RUBoost-based data science framework to evaluate the influences of the same component parameters on cell capacity is also carried out for LTO-based battery.

For LTO-based battery cell capacity, its classification error would converge to 0.06 after using 92 decision trees. The Binder type provides the lowest value of Gini importance while LTOw, Binderw, and C65w result in the three most important

Fig. 3.27 Confusion matrix for LFP capacity case. **a** All feature variables; **b** reduced feature variables, reprinted from [44], with permission from IEEE

parameters with weights over 0.08, as illustrated in Fig. 3.28. Based upon its heat map in Fig. 3.29, the pair of LTOw and CNFw gives the highest PMOA but is still less than 0.5, which implies that the correlations are small for the classification of LTO-based battery.

Interestingly, according to the CMs of both all and reduced material component parameter cases in Fig. 3.30, the classification results of these two cases are similar with a *microF*1 of 94.4%. These facts signify that a satisfactory cell capacity classification could be obtained for LTO-based battery. On the other hand, the component parameters of LTOw, Binderw, and C65w are enough to accurately classify the LTO-based battery cell capacity through using the derived RUBoost-based data science framework.

In summary, the class imbalance problem is easy to happen during battery manufacturing process and would further affect the classification performance of manufactured cell capacity. The proposed RUBoost-based data science framework presents

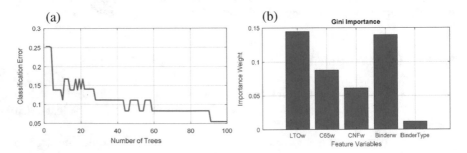

Fig. 3.28 Classification error and Gini importance for LTO capacity case. **a** Classification error versus M; **b** Gini importance, reprinted from [44], with permission from IEEE

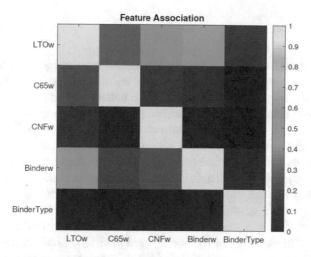

Fig. 3.29 Feature association for LTO capacity classification case, reprinted from [44], with permission from IEEE

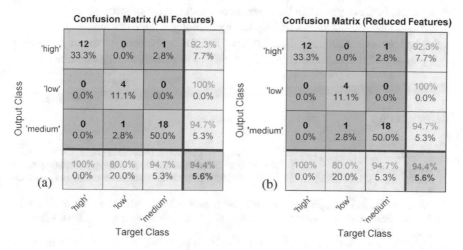

Fig. 3.30 Confusion matrix for LTO capacity case. **a** All feature variables; **b** reduced feature variables, reprinted from [44], with permission from IEEE

the superiorities in terms of accuracy, interpretability for both feature importance and correlations, data-driven nature, and the ability to handle the class imbalance issue. When other battery manufacturing data such as the mixing speed, kneading intensity, temperature, and pressure are available, it has a good potential in the reliable multi-classification and sensitivity analyses for these process parameters to benefit smarter battery manufacturing.

3.5   Summary

This chapter describes the data science-based battery manufacturing management, the initial and key stage during battery full-lifespan. An overview of battery manufacturing is first introduced by dividing it into battery electrode manufacturing and battery cell manufacturing. Then a framework for using data science tools to manage battery manufacturing is described, followed by the comparisons of various popular machine learning technologies used in battery manufacturing management in terms of their merits and drawbacks. For battery electrode manufacturing, two data science-based case studies through deriving the GPR regression model and RF classification model to predict manufactured battery electrode properties (mass loading and porosity) and perform sensitivity analyses of strong-coupled manufacturing parameters are given. For battery cell manufacturing, another two data science-based case studies by designing SVR and RUBoost-based models to predict/classify manufactured cell capacities and analyse related manufacturing parameters are described. Illustrative results indicate that designing suitable data science tools could accurately capture battery product properties prior to its manufacturing. Furthermore, it could automatically explain the interactions and effects of strong-coupled manufacturing parameters. This could help to significantly reduce the monitoring burden of the battery manufacturing line. Cheaper and more efficient means of producing high-performance batteries can be also identified to benefit battery manufacturers.

References

1. Kwade A, Haselrieder W, Leithoff R, Modlinger A, Dietrich F, Droeder K (2018) Current status and challenges for automotive battery production technologies. Nat Energy 3(4):290–300
2. Kendrick E (2019) Advancements in manufacturing. In: Future lithium-ion batteries, pp 262–289
3. Turetskyy A, Thiede S, Thomitzek M, Von Drachenfels N, Pape T, Herrmann C (2020) Toward data-driven applications in lithium-ion battery cell manufacturing. Energy Technol 8(2):1900136
4. Ng M-F, Zhao J, Yan Q, Conduit GJ, Seh ZW (2020) Predicting the state of charge and health of batteries using data-driven machine learning. Nat Mach Intell 2(3):161–170
5. Aykol M, Herring P, Anapolsky A (2020) Machine learning for continuous innovation in battery technologies. Nat Rev Mater 5(10):725–727
6. Niri MF, Liu K, Apachitei G, Román-Ramírez LA, Lain M, Widanage D, Marco J (2022) Quantifying key factors for optimised manufacturing of Li-ion battery anode and cathode via artificial intelligence. Energy AI 7:100129
7. Wanner J, Weeber M, Birke KP, Sauer A (2019) Quality modelling in battery cell manufacturing using soft sensing and sensor fusion—a review. In: Proceedings of 9th international electric drives production conference (EDPC), SV Veranstaltungen, Germany, Esslingen, 2019, pp 1–9
8. Schnell J, Reinhart G (2016) Quality management for battery production: a quality gate concept. Procedia CIRP 57:568–573
9. Schnell J, Nentwich C, Endres F, Kollenda A, Distel F, Knoche T, Reinhart G (2019) Data mining in lithium-ion battery cell production. J Power Sources 413:360–366
10. Thiede S, Turetskyy A, Kwade A, Kara S, Herrmann C (2019) Data mining in battery production chains towards multi-criterial quality prediction. CIRP Ann 68(1):463–466

11. Hoffmann L, Grathwol J-K, Haselrieder W, Leithoff R, Jansen T, Dilger K, Dröder K, Kwade A, Kurrat M (2020) Capacity distribution of large lithium-ion battery pouch cells in context with pilot production processes. Energy Technol 8(2):1900196
12. Cunha RP, Lombardo T, Primo EN, Franco AA (2020) Artificial intelligence investigation of NMC cathode manufacturing parameters interdependencies. Batteries Supercaps 3(1):60–67
13. Riexinger G, Doppler JP, Haar C, Trierweiler M, Buss A, Schöbel K, Ensling D, Bauernhansl T (2020) Integration of traceability systems in battery production. Procedia CIRP 93:125–130
14. Wessel J, Turetskyy A, Wojahn O, Herrmann C, Thiede S (2020) Tracking and tracing for data mining application in the lithium-ion battery production. Procedia CIRP 93:162–167
15. Knoche T, Surek F, Reinhart G (2016) A process model for the electrolyte filling of lithium-ion batteries. Procedia CIRP 41:405–410
16. Schönemann M, Bockholt H, Thiede S, Kwade A, Herrmann C (2019) Multiscale simulation approach for production systems. Int J Adv Manuf Technol 102(5):1373–1390
17. Kornas T, Knak E, Daub R, Bührer U, Lienemann C, Heimes H, Kampker A, Thiede S, Herrmann C (2019) A multivariate KPI-based method for quality assurance in lithium-ion-battery production. Procedia CIRP 81:75–80
18. Niri MF, Liu K, Apachitei G, Roman-Ramirez L, Lain M, Widanalage D, Marco J (2021) Machine-learning for Li-ion battery capacity prediction in manufacturing process. In: Proceedings of ECS meeting abstracts, p 427
19. Zhang S, Zhang C, Yang Q (2003) Data preparation for data mining. Appl Artif Intell 17(5–6):375–381
20. Liu K, Yang Z, Wang H, Li K (2021) Classifications of lithium-ion battery electrode property based on support vector machine with various kernels. In: Recent advances in sustainable energy and intelligent systems. Springer, Singapore, pp 23–34
21. Emilsson E, Dahllöf L (2019) Lithium-ion vehicle battery production. IVL Swedish Environmental Research Institute, Stockholm, Sweden
22. Liu K, Peng Q, Li K, Chen T (2022) Data-based interpretable modeling for property forecasting and sensitivity analysis of Li-ion battery electrode. Autom Innov 1–13
23. Lenze G, Bockholt H, Schilcher C, Froböse L, Jansen D, Krewer U, Kwade A (2018) Impacts of variations in manufacturing parameters on performance of lithium-ion-batteries. J Electrochem Soc 165(2):A314
24. Mohanty D, Hockaday E, Li J, Hensley D, Daniel C, Wood III D (2016) Effect of electrode manufacturing defects on electrochemical performance of lithium-ion batteries: cognizance of the battery failure sources. J Power Sources 312:70–79
25. Baunach M, Jaiser S, Schmelzle S, Nirschl H, Scharfer P, Schabel W (2016) Delamination behavior of lithium-ion battery anodes: influence of drying temperature during electrode processing. Drying Technol 34(4):462–473
26. Liu K, Wei Z, Yang Z, Li K (2021) Mass load prediction for lithium-ion battery electrode clean production: a machine learning approach. J Clean Prod 289:125159
27. Rasmussen CE, Nickisch H (2010) Gaussian processes for machine learning (GPML) toolbox. J Mach Learn Res 11:3011–3015
28. Liu D, Pang J, Zhou J, Peng Y, Pecht M (2013) Prognostics for state of health estimation of lithium-ion batteries based on combination Gaussian process functional regression. Microelectron Reliab 53(6):832–839
29. Liu K, Hu X, Zhou H, Tong L, Widanalage D, Marco J (2021) Feature analyses and modelling of lithium-ion batteries manufacturing based on random forest classification. IEEE/ASME Trans Mechatron 26(6):2944–2955
30. Cutler A, Cutler DR, Stevens JR (2012) Random forests. In: Ensemble machine learning. Springer, Boston, MA, pp 157–175
31. Liu H, Cocea M (2018) Induction of classification rules by Gini-index based rule generation. Inf Sci 436:227–246
32. Bergstra J, Bengio Y (2012) Random search for hyper-parameter optimization. J Mach Learn Res 13(2)

33. Pfleging W (2018) A review of laser electrode processing for development and manufacturing of lithium-ion batteries. Nanophotonics 7(3):549–573
34. Leithoff R, Fröhlich A, Dröder K (2020) Investigation of the influence of deposition accuracy of electrodes on the electrochemical properties of lithium-ion batteries. Energy Technol 8(2):1900129
35. Schilling A, Wiemers-Meyer S, Winkler V, Nowak S, Hoppe B, Heimes HH, Dröder K, Winter M (2020) Influence of separator material on infiltration rate and wetting behavior of lithium-ion batteries. Energy Technol 8(2):1900078
36. Heins TP, Leithoff R, Schlüter N, Schröder U, Dröder K (2020) Impedance spectroscopic investigation of the impact of erroneous cell assembly on the aging of lithium-ion batteries. Energy Technol 8(2):1900288
37. Francis CF, Kyratzis IL, Best AS (2020) Lithium-Ion battery separators for ionic–liquid electrolytes: a review. Adv Mater 32(18):1904205
38. Weber CJ, Geiger S, Falusi S, Roth M (2014) Material review of Li ion battery separators. In: Proceedings of American Institute of Physics Conference (AIP), TU Bergakademie, Germany, Freiberg, 2014, pp 66–81
39. Wood III DL, Li J, An SJ (2019) Formation challenges of lithium-ion battery manufacturing. Joule 3(12):2884–2888
40. Zhou Y, Su M, Yu X, Zhang Y, Wang J-G, Ren X, Cao R, Xu W, Baer DR, Du Y (2020) Real-time mass spectrometric characterization of the solid–electrolyte interphase of a lithium-ion battery. Nat Nanotechnol 15(3):224–230
41. Niri MF, Liu K, Apachitei G, Ramirez LR, Lain M, Widanage D, Marco J (2021) Machine learning for optimised and clean Li-ion battery manufacturing: revealing the dependency between electrode and cell characteristics. J Clean Prod 324:129272
42. Noble WS (2006) What is a support vector machine? Nat Biotechnol 24(12):1565–1567
43. Rebentrost P, Mohseni M, Lloyd S (2014) Quantum support vector machine for big data classification. Phys Rev Lett 113(13):130503
44. Liu K, Hu X, Meng J, Guerrero JM, Teodorescu R (2021) RUBoost-based ensemble machine learning for electrode quality classification in Li-ion battery manufacturing. IEEE/ASME Trans Mechatron. https://doi.org/10.1109/TMECH.2021.3115997 (in press)
45. Ying C, Qi-Guang M, Jia-Chen L, Lin G (2013) Advance and prospects of AdaBoost algorithm. Acta Automat Sin 39(6):745–758
46. Mounce S, Ellis K, Edwards J, Speight V, Jakomis N, Boxall J (2017) Ensemble decision tree models using RUSBoost for estimating risk of iron failure in drinking water distribution systems. Water Resour Manag 31(5):1575–1589

Chapter 4
Data Science-Based Battery Operation Management I

This chapter focuses on the data science technologies for battery operation management, which is another key and intermediate process in the full-lifespan of battery. After manufacturing, battery would be operated in various applications such as transportation electrification, stationary energy storage and smart grid to supply or absorb the power, where suitable management solutions are necessary to ensure its efficiency, safety, and sustainability. In this context, numerous state-of-the-art data science strategies have been developed to perform efficient management of battery operation.

To systematically illustrate the data science-based strategies for benefitting battery operation management, an overview is first given to introduce several crucial parts of battery operation management, which includes battery operation modelling, battery state estimation, battery lifetime prognostics, battery fault diagnosis, and battery charging. Then the fundamentals of battery operation modelling as well as state estimation are detailed in this chapter, while the latter three operation parts will be described in the next chapter. Besides, case studies of deriving proper data science methods to benefit four crucial state estimations of battery are all presented and analysed.

4.1 Battery Operation Modelling

Establishing a suitable battery model is generally the starting point for battery operation management [1]. Over the years, numerous data science-based battery operation models with different levels of accuracy and complexity have been designed. This section mainly focuses on three typical types of battery operation models including battery electrical model, battery thermal model, and battery coupled model, which are widely adopted to capture battery operational dynamics, as detailed in Fig. 4.1.

© The Author(s) 2022
K. Liu et al., *Data Science-Based Full-Lifespan Management of Lithium-Ion Battery*, Green Energy and Technology,
https://doi.org/10.1007/978-3-031-01340-9_4

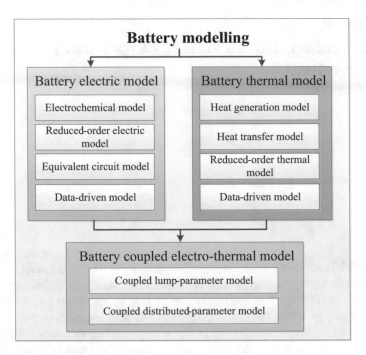

Fig. 4.1 Three typical and widely used battery operation models

4.1.1   Battery Electrical Model

As a fundamental battery operation model, electrical model can be mainly divided into the electrochemical model [2, 3], reduced-order model [4, 5], equivalent circuit model [6, 7] and machine learning model [8–10].

For the electrochemical model of battery, Rahman et al. [2] claim that this type of battery electrical model should own the ability to describe the spatiotemporal dynamics of battery concentration, the electrode potential of each phase, and the Butler–Volmer kinetic for controlling intercalation reactions. Then an electrochemical model is established for describing battery electrochemical behaviours while its parameters are optimized by the particle swarm optimization (PSO) approach. Sung et al. [3] present that the battery electrochemical model is able to provide a highly accurate prediction performance, but it usually needs significant computational efforts for model simulation. Then a model implementation solution is designed to embed this complicated model into the BMS. The main merit of adopting electrochemical model is that a highly accurate description of electrochemical process within a battery could be achieved. However, numerous parameters that reflect battery electrochemistry such as the chemical composition require to be identified by using data science tool, which is actually a big challenge in real battery operation applications. Moreover, many partial differential equations are involved in a battery electrochemical model, bringing a large computational burden to solve them. It should be known that through making proper assumptions, the full-order electro-chemical model could be approximated by the reduced-order model. For instance, after capturing both solid-phase diffusion and electrolyte concentration distribution within battery based on an approximate approach, a simplified physics-based electro-chemical model is derived in [4]. Li et al. [5] simplify the electrochemical model with reduced-order to predict discharging capacities of LiFePO$_4$ battery under different conditions. Although some information would inevitably be lost by using the simpli-fied reduced-order model, this type of electrical model would become more desirable for real operation management of battery. Here, the computational effort becomes much lower by using a reduced-order model, while its corresponding parameters could be identified based on the data from measured battery terminal current and voltage.

For the equivalent circuit model of battery, battery electrical dynamics can be captured by using a combination of circuit components including the resistance, capacity, and voltage sources. Due to the simple structure and relatively small amount of parameters need to be identified, equivalent circuit model has become one of the most popular ones for battery real operation management. Figure 4.2 illustrates a classical framework of battery equivalent circuit model with two resistance–capaci-tance (RC)-networks. Specifically, these RC-networks could reflect battery electrical behaviours such as the charge transfer or diffusion process. Here, the number of RC-networks is generally treated as the order of equivalent circuit model, which requires to be carefully chosen. According to [6], equivalent circuit model with one or two RC-networks could provide satisfactory performance, while higher order of RC-networks becomes unnecessary in many battery operation cases. Nejad et al. [7] provide a critical review for widely adopted equivalent circuit model for Li-ion batteries. Illustrative results indicate that equivalent circuit model with RC-network presents better dynamic performance particular for the prognostics of battery state of charge (SoC) and power.

For the machine learning-based model, various machine learning technologies such as neural networks (NNs) [8, 9] and support vector machine (SVM) [10] have been adopted to derive suitable data science models for capturing battery electrical

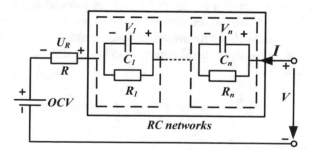

Fig. 4.2 Typical structure of battery equivalent circuit model

dynamics without the requirements of battery prior knowledge. The performance of this type of battery model significantly relies on the experimental data as well as training solutions. To achieve satisfactory prediction accuracy and good generalization performance, experimental data should cover enough battery operation ranges, while the parameters of machine learning models require to be carefully optimized by using suitable training solutions. Moreover, the adaptive data science techniques [11, 12] could be adopted to provide better modelling results.

4.1.2 Battery Thermal Model

Apart from battery electrical dynamics, battery thermal behaviour such as its temperature variation is another key aspect to affect battery operation management because it plays a pivotal role in determining battery performance and service life [13–15]. In this context, different data science models such as heat generation model, heat transfer model, reduced-order thermal model, and machine learning-based model have been proposed to describe battery thermal dynamics. For battery heat generation model, a great deal of solutions are designed to capture battery heat generation, such as activation, concentration, and ohmic loss, which distribute non-uniform within a battery. Three popular ways to assess battery heat generation are described in Eq. (4.1), which have been widely adopted for real battery operation management [16, 17].

$$
\begin{cases}
Q_a = R \cdot I^2 \\
Q_b = I \cdot (V - \text{OCV}) \\
Q_c = I \cdot (V - \text{OCV}) + I \cdot T \cdot \text{dOCV}/\text{d}T
\end{cases}
\tag{4.1}
$$

where R stands for battery internal resistance. I and V represent terminal current and voltage of battery, respectively. OCV means open-circuit voltage of battery. Q_a is the heat generation mainly caused by large currents that across battery internal resistance. Q_b represents heat generation caused by overpotential across RC-networks. Q_c denotes heat generation due to the entropy variation as well as Joule heating.

Besides, for the heat transfer of battery, the convection, conduction, and radiation of heat are the three primary forms within as well as outside a battery [18, 19]. A three-dimensional distributed-parameter heat transfer model is developed by Guo et al. [20] to explore the geometrical currents and heat distribution within a Li-ion battery, as described by,

$$
\frac{\partial \rho C_p T_{3C}}{\partial t} = -\nabla (k_{3C} \nabla T_{3C}) + Q
\tag{4.2}
$$

It can be also expressed by [21],

$$\frac{\partial \rho C_p T_{3C}}{\partial t} = -\frac{\partial}{\partial x}\left(k_x \frac{\partial T_{3C}}{\partial x}\right) - \frac{\partial}{\partial y}\left(k_y \frac{\partial T_{3C}}{\partial y}\right) - \frac{\partial}{\partial z}\left(k_z \frac{\partial T_{3C}}{\partial z}\right) + Q \quad (4.3)$$

where ρ reflects battery density, C_p stands for heat capacity of battery, k_{3C} is a coefficient to reflect battery thermal conductivity (along three dimensions: k_x, k_y, k_z), and Q stands for the heat generation of battery.

Supposing the temperature distribution of a battery within each layer plane is uniform, and only considering one dimension (x, y, z) of battery heat conduction, then a one-dimensional heat conduction thermal model can be simplified as [22],

$$\frac{\partial \rho C_p T_{1C}}{\partial t} = -\frac{\partial}{\partial x}\left(k_x \frac{\partial T_{1C}}{\partial x}\right) + Q \quad (4.4)$$

The three-dimensional heat transfer model is able to describe temperature distribution within a battery, which could be further adopted to detect possible hot spots, especially for high heat generation operations. The one-dimensional heat transfer model is capable of capturing battery temperature gradient along a direction of interest. However, the computational efforts of these heat transfer models are usually too large to be applied in real battery operation management, and they are primarily adopted in offline simulation conditions.

Let heat conduction becomes the only type for heat transfer, heat generation is evenly distributed within a battery, while the temperatures of both battery surface and interior become uniform, then a two-stage battery thermal model [23, 24] that has been widely used in battery operation management is derived as:

$$\begin{cases} C_{q1} \cdot dT_{in}/dt = k_1 \cdot (T_{sh} - T_{in}) + Q \\ C_{q2} \cdot dT_{sh}/dt = k_1 \cdot (T_{in} - T_{sh}) + k_2 \cdot (T_{amb} - T_{sh}) \end{cases} \quad (4.5)$$

where T_{in} is battery internal temperature, while T_{sh} is battery surface temperature; T_{amb} stands for ambient temperature; C_{q1} is battery internal thermal capacity while C_{q2} represents battery surface thermal capacity; k_1 reflects the heat conduction between battery surface and interior, k_2 is the heat conduction between battery surface and ambient temperature.

After defining battery heat generation and transfer parts, numerous battery thermal models with reduced order have been also successfully designed to achieve control purposes for battery operation management [25]. After converting the one-dimensional boundary-value issue into a linear model with low order in the frequency domain, the order of a Li-ion battery thermal model can be decreased, while its temperature prediction could match closely with the results of experiment and three-dimensional finite-element simulations. According to the computational fluid dynamics (CFD) model, a reduced-order state-space thermal model is proposed by using the singular value decomposition method [25], while the similar results as CFD model can be achieved but with much less computational efforts.

4.1.3 Battery Coupled Model

In battery operation applications, there exists strong coupling among different battery dynamics. For example, battery electrical and thermal behaviours are strongly coupled with each other. To better capture battery electrical dynamics (e.g. current, voltage, SoC) and thermal dynamics (e.g. surface and internal temperature), several battery coupled electrothermal models have been proposed, including lump-parameter model and distributed-parameter model [26–28]. For example, a three-dimensional electrothermal model is proposed by Goutam et al. [29] to predict battery SoC and calculate its heat generation. Specifically, this electrothermal model contains a 2D potential distribution model and a 3D temperature distribution model. Then both battery SoC and temperature distribution under constant as well as dynamic currents could be effectively obtained by using this coupled model. In [30], an electrothermal model with decreased order is designed and evaluated by batteries with three different cathode materials. This reduced model is accurate enough for developing quick heating and optimal charging method at low temperature cases. A coupled electrothermal model with three-dimensional is proposed by Basu et al. [31] to analyse the effects of different battery operations such as coolant flow-rates and discharge currents on the variations of battery temperature, further verifying that the contact resistance plays a pivotal role in affecting battery temperature.

4.2 Battery State Estimation

Due to the complex electrochemical characteristics and multi-physics coupling, a trivial emulation of battery operations based on just measured voltage, current, and surface temperature cannot lead to the in-depth understanding or monitoring of operated batteries. In this context, performing accurate estimation of several battery internal states is crucial for advanced battery operation management [32]. This section details the data science-based battery state estimation with a focus on several battery fundamental but important states including the state of charge (SoC), state of power (SoP), state of health (SoH), and joint states estimation.

4.2.1 Battery SoC Estimation

4.2.1.1 Definition of Battery SoC

SoC is a fundamental and critical factor for the operation management of battery, which can be expressed by different formulation forms [33]. SoC generally represents the battery available capacity (C_a) expressed as a percentage of its nominal capacity (C_n) [34]. C_n is the maximum charge amount that can be stored within a battery.

Similar to the fuel vehicle's tank, SoC presents the same functionality as a fuel gauge. Supposing current I is positive and negative for charging and discharging, respectively, a common definition of battery SoC is:

$$\text{SoC}(t) = \text{SoC}(t_0) + \int_{t_0}^{t} \frac{\eta \cdot I(t)}{C_n} \, dt \tag{4.6}$$

where $\text{SoC}(t)$ and $\text{SoC}(t_0)$, respectively, denote battery SoC values at time point t and initial time point t_0; η represents coulombic efficiency to reflect the ratio of fully discharged energy to charged energy needed to recover the original capacity.

On the other hand, from the battery electrochemical side, SoC could represent the charge contained in electrode particles. Specifically, the variation of battery SoC is able to reflect the distribution of lithium concentration within electrode particles. Due to the amount of available charge highly relies on the amount of lithium stored in electrodes, SoC can be directly obtained by considering the mean lithium concentration $\overline{C}_s$ as:

$$\text{SoC}(t) = \frac{\overline{C}_s(t) - C_{s,\min}}{C_{s,\max} - C_{s,\min}} \tag{4.7}$$

where $\overline{C}_s(t)$ represents the mean surface Li-ion concentration at time point t, and $C_{s,\min}$ and $C_{s,\max}$ stand for the surface Li-ion concentrations when fully charging and discharging a battery, respectively.

The operation management of battery requires accurate SoC information to indicate the remaining available energy within a battery during its operations. In the laboratory conditions, based upon a known initial SoC value, the referenced battery SoC is generally obtained through a well-controlled coulomb counting approach to accumulate charge transferred [35]. However, complex battery electrochemical dynamics and strongly coupled characteristics make battery SoC is difficult to be measured directly in real-world applications. In this context, reliable battery SoC estimation in real time is a critical part of battery operation management, thus attracting considerable data science research efforts.

4.2.1.2 Data Science-Based SoC Estimation Methods

To date, different data science-based methods were designed to achieve reasonable SoC estimation for battery operation management in the literature. These data science-based methods could be divided into three main categories including the direct calculation method, model-based method, and machine learning method, as shown in Fig. 4.3.

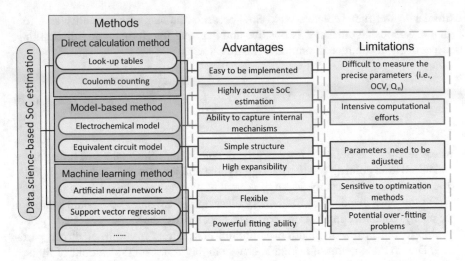

Fig. 4.3 Data science-based battery SoC estimation methods in terms of merits and limitations, reprinted from [32], with permission from Elsevier

For data science direct calculation method, two common solutions are noteworthy. First, as there exist obvious mapping relations between battery SoC and some battery direct factors such as the open-circuit voltage (OCV) and impedance, after obtaining the data of these battery factors, the battery SoC could be inferred by the predefined lookup tables that describe such a relation [36]. Besides, based upon the obtained data of battery nominal capacity and exact current profiles, the variation of battery SoC could be conveniently obtained by the coulomb counting approach. One obvious superiority of these two methods is the easy way to be implemented for battery SoC estimation. However, as battery capacity would inevitably vary with different ageing levels, and OCV needs to be obtained after a rest period, precisely online measuring these data is still a daunting challenge during battery operation management. In the light of this, attempts have been made to estimate battery SoC through other data science supporters such as model-based methods.

For the model-based method, according to a proper model such as dynamics equations, different estimation strategies together with easily collect current or voltage data would be employed to estimate battery SoC. One popular model type is the battery electrochemical model (EM) as battery internal mechanisms such as kinetic and charge transfers within a battery could be described by it, further leading to an accurate SoC estimation of battery. However, as EM-based methods involve many parameters and partial differential equations, it generally requires a large computational burden to implement them. In order to facilitate a real-time application, suitable simplifications are always required. Another widely used model type is the battery equivalent circuit model (ECM) which adopts electrical circuit components to emulate battery dynamics. Due to a relatively simple structure and reasonable expansibility, ECM becomes a promising tool for battery real-time SoC estimation. However, considering the ECMs' parameters would vary over time, using the invariant parameters under different temperature, SoC or ageing levels could cause large estimation error [37]. Great efforts are required to periodically recalibrate ECMs' parameters, further ensuring their extensibility. Besides, after obtaining different battery data in real applications, it is vital to develop proper solutions such as the joint parameter/SoC estimation tool for adjusting model parameters adaptively.

The machine learning method, free of understanding battery electrochemical mechanisms, has also been widely adopted to estimate battery SoC. Due to the potential merits such as being flexible and highly nonlinear fitting ability, various techniques including the neural network (NN) [38] and support vector machine (SVM) [39] have been utilized to estimate battery SoC. However, these data science solutions are very sensitive to their optimization strategies and the quality of collected data. In addition, overfitting issues would happen if improper training modes are utilized.

4.2.1.3 Case Study: Battery SoC Estimation with RLS and EKF

In this subsection, we will introduce a data science-based SoC estimation method for Li-ion battery with recursive least square (RLS) and extended Kalman filter (EKF). As the most widely used model-based method, the RLS-EKF integrates the RLS-based online model identification and the EKF-based SoC observation in a dual sequential framework.

The estimation accuracy of this method largely depends on the accuracy of the model. However, a model with a higher accuracy requires more parameters to be identified, that is to say, a higher computational cost is required. Generally, a battery model suitable for the real-time application has a simple topology while capturing the major dynamics of Li-ion battery. In the light of this, this case study uses the Thevenin model to verify the RLS-EKF algorithm. As is shown in Fig. 4.4, R_0 is the ohmic resistance. The single RC branch is used to simulate the polarization effects due to passivation layers on the electrodes, charge transfer between electrode and electrolyte, diffusion, migration, and convection processes. The dynamics of the Thevenin model in use is written as:

$$\begin{cases} U_t(t) = U_{\mathrm{OCV}}(\mathrm{SoC}(t)) - i(t)R_0 - U_1(t) \\ \dot{U}_1(t) = -\frac{U_1(t)}{R_1 C_1} + \frac{i(t)}{C_1} \end{cases} \tag{4.8}$$

Further discretizing the above formula can obtain:

$$\begin{cases} U_t(k) = U_{\mathrm{OCV}}(\mathrm{SoC}(k)) - i(k)R_0 - U_1(k) \\ U_1(k+1) = e^{-\frac{\Delta t}{R_1 C_1}} U_1(k) + R_1(1 - e^{-\frac{\Delta t}{R_1 C_1}})i(k) \end{cases} \tag{4.9}$$

The OCV is a nonlinear function of SoC which should be calibrated accurately to adopt the model-based estimation. In order to identify the functional relationship between OCV-SoC, it is necessary to obtain the OCV-SoC curve through the OCV experiment, fit the experimental curve through the empirical function, identify the

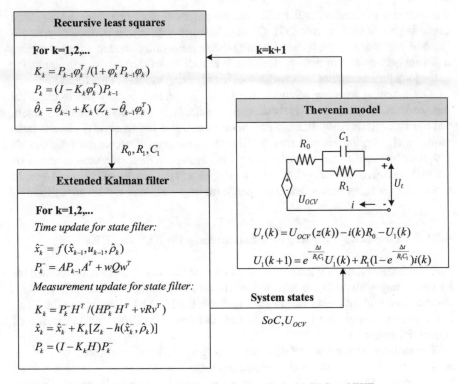

Fig. 4.4 Data science-based battery SoC estimation method with RLS and EKF

parameters of the given function through the parameter identification method, and finally get the OCV-SoC function relationship. It is usually possible to use higher-order polynomial approximation as:

$$U_{\text{OCV}}(z) = k_0 + k_1 z + k_2 z^2 + \cdots + k_n z^n \qquad (4.10)$$

where z denotes the SoC, k_n ($n = 1, 2, \ldots, n_f$) are the polynomial coefficients, and n_f is the polynomial order which is selected to be 5 in this case.

Since the sampling of current and voltage signals is discrete in the actual battery system, the continuous-time model of SoC should also be discretized:

$$\text{SoC}(k + 1) = \text{SoC}(k) - \eta \frac{i(k)\Delta t}{Q} \qquad (4.11)$$

The adoption of RLS-based online model identification necessitates formulating a regression model. To achieve this, a new variable is defined as:

$$\frac{U_t(s) - U_{\text{OCV}}(s)}{i(s)} = \left(-R_0 + \frac{R_1}{1 + R_1 C_1 s}\right) \qquad (4.12)$$

By applying the bilinear transform, Eq. (4.12) can be transformed to:

$$\frac{U_t(z) - U_{OCV}(z)}{i(z)} = \frac{c_2 + c_3}{1 - c_1 z^{-1}} \tag{4.13}$$

where

$$c_1 = \frac{2R_1C_1 - \Delta t}{2R_1C_1 + \Delta t}$$

$$c_2 = \frac{(R_0 + R_1)\Delta t + 2R_0R_1C_1}{-(\Delta t + 2R_1C_1)}$$

$$c_3 = \frac{(R_0 + R_1)\Delta t - 2R_0R_1C_1}{-(\Delta t + 2R_1C_1)}$$

Then the discrete-time expression of the system can be written as:

$$U_{q,k} = \theta_k \phi_k^T \tag{4.14}$$

where $U_{q,k} = U_{t,k}$, $\theta_k = [(1 - c_1)U_{OC,k}, c_1, c_2, c_3]$, $\phi_k = [1, U_{t,k-1}, i_k, i_{k-1}]$. The regression model is then identifiable with the RLS method. A forgetting factor (λ) is used to give more weight to the recently obtained data while discounting the contribution of historical data. Once the model parameters are obtained, the system states including polarization voltage and SoC are estimated by using EKF. Choose SoC and U_1 as the state variable, electric current as input quantity, and terminal voltage (U_t) as output quantity. The equation can be transformed as:

$$\begin{cases} \begin{bmatrix} U_1(k) \\ SoC(k) \end{bmatrix} = \begin{bmatrix} e^{\left(-\frac{\Delta t}{R_1 C_1}\right)} & 0 \\ 0 & 1 \end{bmatrix} \begin{bmatrix} U_1(k-1) \\ SoC(k-1) \end{bmatrix} + \begin{bmatrix} R_1\left(1 - e^{\left(-\frac{\Delta t}{R_1 C_1}\right)}\right) \\ -\eta \Delta t / Q_N \end{bmatrix} i(k-1) \\ \quad + w_{k-1} \\ U_t(k) = f[SoC(k)] - U_1(k) + R_0 i(k) + v_k \end{cases} \tag{4.15}$$

where w_{k-1} is the random process noise sequence, and v_k is the random observation noise sequence. Since $f[SoC(k)]$ is nonlinear with respect to SoC, the formula is expanded into a Taylor series around the optimal prediction estimate to linearize. The RLS-based model identification and EKF-based SoC estimation are integrated in a dual sequential framework as shown in Fig. 4.4.

In this case, A123 ANR26650 M1-B batteries with a nominal capacity of 2.5 Ah are selected to verify the algorithm performance. In order to obtain the maximum usable capacity of the battery, a constant volume experiment must be carried out at a room temperature of 25 °C. In detail, a constant-current-constant-voltage method with 1C rate is used to charge the batteries until the voltage reaches 3.6 V and then current drops to 0.05C. A constant-current method with 1C rate is used to discharge cells until the voltage drops to 2.0 V. The battery OCV has a monotonic mapping relationship with the SoC. Determining the OCV-SoC mapping rule is of great significance for improving the accuracy of battery modelling and SoC state estimation. Therefore, after the constant volume test, it is necessary to calibrate the OCV-SoC curve. Specifically, taking the charging OCV as an example, if 10% SoC point is used as the test point, battery is discharged to the cut-off voltage 2.75 V with a 1C current at a constant current, and then left for 2 h as the voltage of SoC = 0%. 1C rate is utilized to charge the battery in cross-current. The cut-off condition is that the charging time reaches 6 min, and after 2 h left, the terminal voltage is recorded as the OCV corresponding to the current SoC, and so on, recording 10%, 20%, ..., 90% OCV. The ECM shown in Fig. 4.4 is built by using MATLAB Simulink. The model parameters are all defined according to the battery testing result. An enhanced UDDS, DST, and FUDS profile shown in Fig. 4.5 is used to verify the feasibility of RLS and EKF algorithms.

As shown in Fig. 4.6, the blue line is the real SoC value, and the red line is the SoC value estimated by the RLS-EKF algorithm. The RMSE and the MAE under different working conditions are shown in Table 4.1.

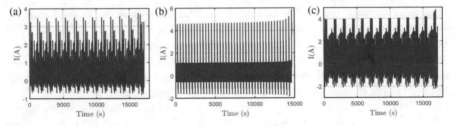

Fig. 4.5 Load profile for testing: **a** an enhanced UDDS profile; **b** an enhanced DST profile; **c** an enhanced FUDS profile

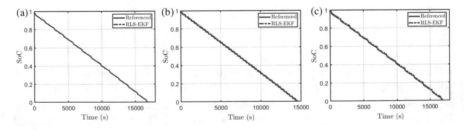

Fig. 4.6 SoC estimation result: **a** UDDS; **b** DST; **c** FUDS

Table 4.1 Performance of RLS-EKF under different working conditions

	UDDS (%)	DST (%)	FUDS (%)
RMSE	0.27	0.87	1.04
MAE	0.75	1.23	1.33

4.2.2 Battery SoP Estimation

4.2.2.1 Definition of Battery SoP

State of power (SoP) is another critical factor for battery operation management and usually utilized to reflect the available power that a battery could supply or absorb over a short time horizon [40]. In theory, battery SoP could be viewed as a result of threshold current and responded voltage, while different operation constraints also need to be explicitly considered. Supposing that discharging power is positive while charging power is negative, battery SoP can be generally expressed by [41]:

$$\begin{cases} \text{SoP}^c(t) = \max\left(P_{\min}, V(t + \Delta t) \cdot I^c_{\min}\right) \\ \text{SoP}^d(t) = \min\left(P_{\max}, V(t + \Delta t) \cdot I^d_{\max}\right) \\ \text{Subject to operation constraints} \end{cases} \tag{4.16}$$

where $\text{SoP}^c(t)$ and $\text{SoP}^d(t)$ denote battery charging and discharging SoPs at time point t, respectively, $P_{\min}$ and $P_{\max}$ represent the minimum and maximum limitations of battery power, Δt is the specific future time period, $V(t + \Delta t)$ stands for battery terminal voltage at time point $t + \Delta t$, and $I^c_{\min}$ and $I^d_{\max}$ are minimum continuous charging current as well as maximum continuous discharging current from time point t to $t + \Delta t$, respectively. $I^c_{\min}$ and $I^d_{\max}$ require to be obtained under the cases of battery operation constraints are not exceeded. These operation constraints usually contain battery voltage, current, SoC, and sometimes temperature.

For the simulation applications, battery SoP reference is generally obtained by the high-fidelity battery model with the consideration of different operation constraints [42]. In the laboratory conditions, battery SoP could be decided through well-designed pulse tests with the consideration of some modified current rate and duration time. For real EV applications, due to the energy-flow management such as the power split and battery charging during regenerative braking highly depends on the available power of battery, reliable battery SoP estimation could benefit not only the regulation of vehicular power flow but also the optimization of overall powertrain efficiency. Moreover, for battery itself, knowing its future SoP could benefit fast charging mode and battery performance. In this context, it is vital to design effective data science strategy for reliable battery SoP estimation that takes the highly nonlinear dynamics and different operation constraints of battery into account.

4.2.2.2 Data Science-Based SoP Estimation Methods

Battery SoP estimation studies are relatively scarce compared with battery SoC estimation methods involving a plethora of research. According to a systematic review in [40], data science-based SoP estimation methods can be mainly divided into two categories, as shown in Fig. 4.7.

For characteristic map (CtM)-based method, a static interdependence between battery SoP and other state variables such as SoC, temperature, voltage, and power pulse duration is established offline. To further enhance the estimation performance of CtM-based method, the difference between measured battery power and estimated SoP value is calculated. Then the reference points within CtM could be adapted in the conditions of a huge deviation that appears [43]. CtM-based method could be readily implemented, owing to its straightforward treatment. However, several issues are still not addressed thoroughly: First, past and current battery information is difficult to be considered in CtMs. As battery power dynamics strongly relies on its operation condition, the accuracy of battery SoP estimation will be thus influenced severely. Second, in order to construct a high-performance CtM under different battery operating conditions, a large amount of information requires to be stored by the multi-dimensional forms, further leading to a large computational burden on the micro-controller. In this context, the online SoP estimation method through deriving suitable battery model with various computational levels is explored and exploited.

For model-based approach, ECM and its variants are usually adopted to estimate battery SoP. After formulating ECM with a discrete-time state-space form, various solutions such as Kalman filter [44] and least square-based approach [42] have been adopted to derive reasonable SoP estimators. To guarantee the estimation accuracy of SoP, an ECM that could not only describe battery overall dynamics but also presents proper structure and parameters becomes necessary. In this context,

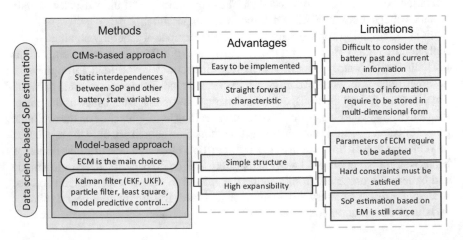

Fig. 4.7 Data science-based battery SoP estimation methods in terms of merits and limitations, reprinted from [32], with permission from Elsevier

ECM's parameters need to be adapted under real-time data from different operating SoC, temperature, and ageing states. In addition, to ensure battery operated safely, the operating constraints of current, voltage, SoC, and/or even internal temperature are required to be satisfied during battery SoP estimation. In addition, ECM-based approach is difficult to depict battery's inside electrochemical process, further leading to poor generalization. Unfortunately, EM-based approach is still scarce in battery SoP estimation domain.

4.2.2.3 Case Study: Battery SoP Estimation with Multi-constrained Dynamic Method

In this subsection, we will introduce a data science-based SoP estimation method for Li-ion battery with the multi-constrained dynamic method. This method comprehensively considers multiple constraint variables, such as terminal voltage, current, SoC, etc., to predict battery SoP in real-time. At the same time, the influence of dynamic response characteristics such as electrochemical kinetics, thermodynamics, and hysteresis effects on the SoP prediction results are comprehensively considered.

As mentioned above, the peak power capability of Li-ion battery is affected by the maximum charge and discharge current, the maximum and minimum cut-off voltage, the remaining available capacity of the battery, etc. In order to estimate the peak power capability accurately, there are multiple constraints (voltage, current, SoC, rated power) that should be taken into consideration, which can be expressed as follows:

$$\begin{cases} U_{min} < U < U_{max} \\ I_{min}^{chg} < I < I_{max}^{dis} \\ SoC_{min} < SoC < SoC_{max} \\ P_{min}^{chg} < P < P_{max}^{dis} \end{cases} \tag{4.17}$$

where U, I, SoC, P represent the battery's terminal voltage, current, SoC, and power. For the battery charge and discharge process, the peak power capability of Li-ion battery can be calculated as follows:

$$\begin{cases} P_{min}^{chg} = \max(P_{min}, UI) \\ P_{max}^{dis} = \min(P_{max}, UI) \end{cases} \tag{4.18}$$

Further, the maximum discharge current and minimum charge current of the battery need to meet the following conditions:

$$\begin{cases} I_{min}^{chg} = \max(I_{min}, I_{min}^{chg,U}, I_{min}^{chg,SoC}) \\ I_{max}^{dis} = \min(I_{max}, I_{max}^{dis,U}, I_{max}^{dis,SoC}) \end{cases} \tag{4.19}$$

Based on the combination of the three constraints, the peak power capability of the battery is finally expressed as follows:

$$\begin{cases} P_{\min}^{\text{chg}} = \max(P_{\min}, P_{\min}^{\text{chg},U}, P_{\min}^{\text{chg},C}, P_{\min}^{\text{chg},\text{SoC}}) \\ P_{\max}^{\text{dis}} = \min(P_{\max}, P_{\max}^{\text{dis},U}, P_{\max}^{\text{dis},C}, P_{\max}^{\text{dis},\text{SoC}}) \end{cases} \Rightarrow \begin{cases} P_{\min}^{\text{chg}} = \max(P_{\min}, U(I_{\min}^{\text{chg}})I_{\min}^{\text{chg}}) \\ P_{\max}^{\text{dis}} = \min(P_{\max}, U(I_{\max}^{\text{dis}})I_{\max}^{\text{dis}}) \end{cases}$$

(4.20)

This case continues to use Thevenin equivalent circuit model to estimate SoP. Similarly, A123 ANR26650 M1-B batteries with a nominal capacity of 2.5 Ah are used to verify the multi-constraint algorithm. In particular, by consulting the battery's user manual, it is necessary to pay attention to the battery-related limit parameters. As shown in Table 4.2, the parameters that need to be paid attention to in SoP estimation are listed. The discharge current is artificially specified as positive.

An enhanced DST profile shown in Fig. 4.8a is used to verify the algorithms. In order to accurately obtain the peak current under the SoC constraint, the RLS-EKF algorithm is first used to estimate battery SoC. Parameter identification and SoC estimation have been introduced in detail in Sect. 4.2.1.3 and will not be repeated here. As shown in Fig. 4.8b, the root-mean-square error (RMSE) is 0.87%, while the mean-absolute error (MAE) is 1.23%.

Next, derive detailed expressions of peak discharge and charge currents under SOC and voltage constraints. The load current is assumed to be constant between the k sampling time and the $(k + L)$ sampling time, where L represents the prediction time horizon. Under the excitation of peak discharge current, the terminal voltage would drop to the lower cut-off voltage, so that the following equation can be drawn:

Table 4.2 Upper and lower cut-off thresholds for SoC, voltage, and current

	SoC (%)	Voltage (V)	Current (A)
Maximum	96	3.6	120
Minimum	5	2	−25

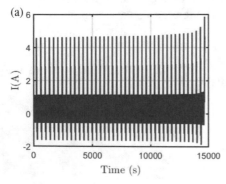
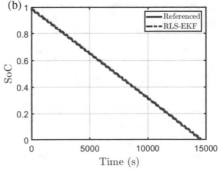

Fig. 4.8 SoC estimation: **a** an enhanced DST profile; **b** SoC estimation results

$$U_{t,\min} = U_{\text{OCV}}(k+L) - U_1(k+L) - I_{L,\max}^{\text{dis,volt}} R_0 \qquad (4.21)$$

The polarization voltage at the $(k+L)$ can be expressed with the battery model as:

$$U_1(k+L) = e^{\frac{-L\Delta t}{R_1 C_1}} U_1(k) + (1 - e^{\frac{-L\Delta t}{R_1 C_1}}) R_1 \sum_{j=1}^{L} e^{\frac{-(j-1)\Delta t}{R_1 C_1}} I_{L,\max}^{\text{dis,volt}} \qquad (4.22)$$

In order to derive $U_{\text{OCV}}(k+L)$, the SoC recurrent relationship is defined as:

$$\text{SoC}(k+L) = \text{SoC}(k) - \frac{\eta I_{L,\max}^{\text{dis,volt}} L \Delta t}{Q} \qquad (4.23)$$

So that,

$$U_{\text{OCV}}(k+L) = U_{\text{OCV}}(k)\left(\text{SoC}(k) - \frac{\eta I_{L,\max}^{\text{dis,volt}} L \Delta t}{Q} \right)$$

$$\approx U_{\text{OCV}}(k) - \frac{\eta I_{L,\max}^{\text{dis,volt}} L \Delta t}{Q} \frac{\partial U_{\text{OCV}}}{\partial \text{SoC}}\Big|_{\text{SoC}(k)} \qquad (4.24)$$

The final expression of the voltage-constrained peak discharge current is:

$$I_{L,\max}^{\text{dis,volt}}(k) = \frac{U_{\text{OCV}}(k) - e^{\frac{-L\Delta t}{R_1 C_1}} U_1(k) - U_{t,\min}}{\frac{\eta L \Delta t}{Q} \frac{\partial U_{\text{OCV}}}{\partial \text{SoC}}\big|_{\text{SoC}(k)} + \left(1 - e^{\frac{-L\Delta t}{R_1 C_1}}\right) R_1 \sum_{j=1}^{L} e^{\frac{-(j-1)\Delta t}{R_1 C_1}} + R_0} \qquad (4.25)$$

The voltage-constrained peak charge current is:

$$I_{L,\min}^{\text{dis,volt}}(k) = \frac{U_{\text{OCV}}(k) - e^{\frac{-L\Delta t}{R_1 C_1}} U_1(k) - U_{t,\max}}{\frac{\eta L \Delta t}{Q} \frac{\partial U_{\text{OCV}}}{\partial \text{SoC}}\big|_{\text{SoC}(k)} + \left(1 - e^{\frac{-L\Delta t}{R_1 C_1}}\right) R_1 \sum_{j=1}^{L} e^{\frac{-(j-1)\Delta t}{R_1 C_1}} + R_0} \qquad (4.26)$$

SoC has to be maintained within a certain range to improve the battery's efficiency and extend the calendar life. The peak discharge and charge current constrained by SoC limit can be derived as:

$$\begin{cases} I_{L,\max}^{\text{dis,SoC}}(k) = \frac{Q(\text{SoC}(k) - \text{SoC}_{\min})}{\eta L \Delta t} \\ I_{L,\max}^{\text{chg,SoC}}(k) = \frac{Q(\text{SoC}(k) - \text{SoC}_{\max})}{\eta L \Delta t} \end{cases} \qquad (4.27)$$

So that multi-constrained peak discharge and charge currents are:

$$\begin{cases} I_{L,\max}^{\text{dis}}(k) = \min\left\{ I_{L,\max}^{\text{dis,volt}}(k),\ I_{L,\max}^{\text{dis,SoC}}(k),\ I_{L,\max}^{\text{dis,current}}(k) \right\} \\ I_{L,\min}^{\text{chg}}(k) = \max\left\{ I_{L,\min}^{\text{chg,volt}}(k),\ I_{L,\min}^{\text{chg,SoC}}(k),\ I_{L,\min}^{\text{chg,current}}(k) \right\} \end{cases} \tag{4.28}$$

After the peak discharge/charge current is determined, the discharge/charge voltage during the prediction time horizon can be derived as:

$$\begin{cases} U_t(k+i) = U_{\text{OCV}}(k) - e^{\frac{-i\Delta t}{R_1 C_1}} U_1(k) \\ \quad - \left[\frac{\eta i \Delta t}{Q} \frac{\partial U_{\text{OCV}}}{\partial \text{SoC}}_{\text{SoC}(k)} + (1 - e^{\frac{-\Delta t}{R_1 C_1}}) R_1 \sum_{j=1}^{i} e^{\frac{-(j-1)\Delta t}{R_1 C_1}} + R_0 \right] I_{L,\max}^{\text{dis}}(k) \\ U_t(k+i) = U_{\text{OCV}}(k) - e^{\frac{-i\Delta t}{R_1 C_1}} U_1(k) \\ \quad - \left[\frac{\eta i \Delta t}{Q} \frac{\partial U_{\text{OCV}}}{\partial \text{SoC}}_{\text{SoC}(k)} + (1 - e^{\frac{-\Delta t}{R_1 C_1}}) R_1 \sum_{j=1}^{i} e^{\frac{-(j-1)\Delta t}{R_1 C_1}} + R_0 \right] I_{L,\max}^{\text{chg}}(k) \end{cases} \tag{4.29}$$

Based on the above derivations, the power sequence under the peak current for the whole prediction time horizon can be expressed as:

$$\begin{cases} P_{\text{dis}}(k+i) = I_{L,\max}^{\text{dis}}(k) U_t(k+i) \\ P_{\text{chg}}(k+i) = I_{L,\min}^{\text{chg}}(k) U_t(k+i) \end{cases} \tag{4.30}$$

where $i = 1, 2, \ldots, L$.

And the final expression of peak power with multi-constrained algorithm can be drawn as:

$$\begin{cases} P_{\text{peak}}^{\text{dis}}(k) = \min_{i=1,2,\ldots,L} [P_{\text{dis}}(k+i)] \\ P_{\text{peak}}^{\text{chg}}(k) = \max_{i=1,2,\ldots,L} [P_{\text{chg}}(k+i)] \end{cases} \tag{4.31}$$

Here, this case choose the charging and discharging time as 10 s to verify the multi-constraint algorithm. As shown in Fig. 4.9, the change curve of peak charge and discharge current value under different constraint conditions (voltage, SoC, and current) and the change curve of peak charge and discharge current value under multiple constraint conditions are given.

According to the value of the above-mentioned current, the terminal voltage is further calculated, and then the peak charge–discharge common rate is calculated. The result of SoP estimation is shown in Fig. 4.10.

4.2.3   Battery SoH Estimation

4.2.3.1   Definition of Battery SoH

Battery would inevitably experience gradual performance fading during its lifetime, owing to its side reaction [45]. In general, battery SoH could be described by its

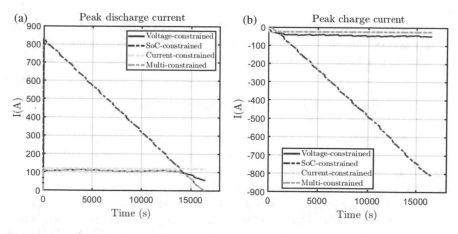

Fig. 4.9 **a** Peak discharge current; **b** peak charge current

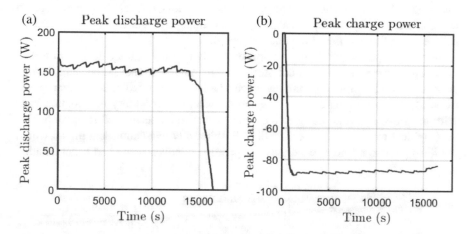

Fig. 4.10 SoP estimation results: **a** peak discharge power; **b** peak charge power

capacity or internal resistance status as:

$$\begin{cases} \text{SoH}_C = \frac{C_a}{C_n} \times 100\% \\ \text{SoH}_R = \frac{R_a - R_r}{R_r} \times 100\% \end{cases} \tag{4.32}$$

where C_a denotes battery actual capacity and C_n is the nominal capacity, and R_a reflects battery actual internal resistance and R_r is the rated internal resistance.

In real applications such as EVs, a 20% capacity degradation and 100% internal resistance increase are generally considered as the end-of-life (EoL) of a battery. In this context, SoH becomes a key factor to underline effective, safe operation management of battery [46]. As it is difficult to directly measure battery capacity and internal resistance with commercially available sensors in real applications, online battery SoH estimation based on the low-cost suite of sensors is crucial for obtaining accurate battery SoH information.

4.2.3.2   Data Science-Based SoH Estimation Methods

A great deal of efforts based on data science techniques has been done for battery SoH estimation, which could be roughly divided into four categories including the physics-based model, empirical model, differential voltage analysis (DVA)/incremental capacity analysis (ICA)-based method [47], and machine learning method. A schematic of available data science-based battery SoH estimation methods is illustrated in Fig. 4.11. Physics-based model adopts partial differential equations (PDEs) to describe battery dynamics of internal physicochemical reactions that are highly related to battery ageing dynamics. This type of model is able to provide clear physical meaning and highly accurate performance. Nevertheless, it faces some challenges in terms of simplifying model and identifying its numerous parameters before it can be fully eligible for real-time implementations [48].

For the empirical model-based approach, after fitting battery degradation data under specific conditions, it could present a light computational burden and provide acceptable SoH estimation accuracy when a battery is operated under similar conditions as the training case [49]. In general, systematical battery degradation tests with laborious and time-consumed efforts require to be performed for establishing an empirical model for SoH estimation. Besides, a derived empirical model would exhibit a poor robustness to the unseen operating conditions and bad generalization

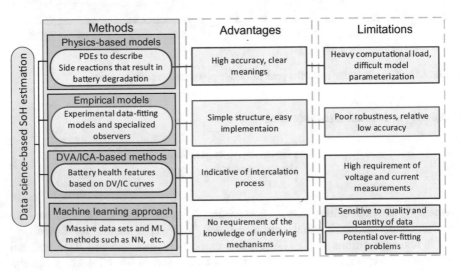

Fig. 4.11 Data science-based battery SoH estimation methods in terms of merits and limitations, reprinted from [32], with permission from Elsevier

ability to batteries with different chemistry or even the dis-similar batch of same chemistry. Therefore, regular model recalibration is vital to increase the related time and cost for developing model. On the other hand, with the rapid development of advanced embedded systems with light computation effort, the physics-based SoH estimation method could be utilized in real battery operation management in future. Then the corresponding simplification and parameterization solutions could become a focus in this direction.

Besides, ICA is also an efficient data science tool to estimate battery SoH [50]. According to the differentiation of charged capacity over battery voltage in the conditions of constant-current charging, the voltage plateaus on battery voltage curve could be transformed into easily identifiable peaks of the IC curve. In this context, the peak position, amplitude, and envelope area of IC curves at different cycles could be utilized to estimate battery SoH [51]. Through using the signal filtering technologies to procure smooth IC curves, the SoH estimation result can be compromised as the peak amplitude is significantly sensitive to the measurement noise. In addition, the voltage range of the voltage curve should cover the voltage corresponding to the peak of the IC curve, which may reduce its feasibility in actual implementation.

Due to the superiority of the mechanism-free nature, advanced machine learning methods such as support vector machine (SVM) and Gaussian process regression (GPR) also become popular for battery SoH estimation [52]. First, a professional battery test that includes all SOH impact factors is carried out, and then the battery SOH model will be synthesized through using machine learning to map these impact factors to the battery SOH. However, the effectiveness of machine learning-based methods largely depends on both the quality and quantity of test data, and the derived models are often affected by the intensity of heavy calculations.

4.2.3.3 Case Study: Battery SoH Estimation with Optimized Partial Voltage Profile

In this subsection, we will introduce a data science-based SoH estimation method for Li-ion battery with optimized partial charging voltage profiles [53]. With a certain amount of dataset from the battery cells, non-dominated sorting genetic algorithm II (NSGA-II) is applied to automatically select the optimal multiple voltage ranges for battery SoH estimation. We can then directly calculate the battery capacity according to the optimized charging voltage profiles.

Normally, the discharging profile of the battery is determined by the load. The charging profile is usually a constant-current constant-voltage (CCCV) process, which is a relatively fixed procedure. Thus, the partial voltage profiles are selected to derive the SoH information in the proposed method.

The battery SoC is defined as,

$$\text{SoC} = \frac{Q_t}{Q_{av}} \times 100\% \tag{4.33}$$

where Q_t is the energy left in the battery, and Q_{av} is the maximum available battery capacity at present. The energy left can be known by the integration of the current flowing in and out of the battery. Thus, we can obtain the Coulomb counting equation as follows,

$$\text{SoC}(k+1) = \text{SoC}(k) + \frac{\eta \cdot I(k) \cdot T_s}{Q_{av}} \tag{4.34}$$

where T_s is the sampling interval. Considering that the Coulombic efficiency is usually above 99.6% for LiFePO$_4$ battery and NMC cell, η is defined as 100% in the rest of the derivation.

According to Eq. (4.34), the following equation can be obtained,

$$Q_{av} = \frac{\sum_{k=A_1}^{A_2} \eta \cdot I(k) \cdot T_s}{\text{SoC}(A_1) - \text{SoC}(A_2)} \tag{4.35}$$

where A_1 and A_2 are the start and termination of the partial voltage profile, respectively.

Generally, the voltage and current are always monitored by a battery management system to ensure the safety of the battery during battery operation management process. It is possible to find a specific voltage ranges from the battery charging process for SoH estimation. Figure 4.12 shows the voltage curve of an NMC-based battery during the charging process. Thus, if the current between U_{A_1} and U_{A_2} is integrated as $\sum_{k=A_1}^{A_2} \eta \cdot I(k) \cdot T_s$ and the SoC variation $(\text{SoC}(A_1) - \text{SoC}(A_2))$ has already known, the battery capacity can be directly calculated from Eq. (4.35).

Now, we can deduce that a proper voltage range $(U_{A_1} \sim U_{A_2})$ should be chosen before estimating the battery SoH during the degradation process. From [53], we know that arbitrarily choosing a voltage profile may not always receive the same good accuracy for SoH estimation. Therefore, it is critical to select the optimal voltage range of the battery capacity prediction when Eq. (4.35) is used. In addition,

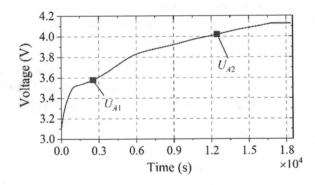

Fig. 4.12 Voltage charge curve of NMC battery, reprinted from [53], with permission from Elsevier

the partial voltage range is also easier to be obtained in the daily usage of the EV compared with the full voltage charging profile. The proposed method is able to effectively compute the Li-ion battery SoH online during the EV charging process.

In this subsection, we plan to propose a methodology to find the optimal voltage range for the battery SoH estimation with data science technique. A single voltage range is firstly considered to predict the battery capacity. Grid search is proposed to optimize a single voltage range with best prediction accuracy. MSE is used to evaluate the accuracy of the estimation and act as the objective function for the grid search optimization. Grid search is an exhaustive searching algorithm, which can select the optimal single voltage range as illustrated in Fig. 4.13. Grid search starts from point A_s and ends at point A_e, the minimal step is S_{min}, and the maximal step is defined as L_{max}. Based on the above definitions, the entire charging voltage curve can be divided into pieces. In this way, grid search can evaluate all the voltage segments and their combinations. The iteration of the grid search will not stop until all the possible voltage ranges are crossed.

Three NMC batteries designed for a market available EV are used to validate the proposed method. The nominal capacity is 63 Ah, and the nominal voltage is 3.7 V, and the voltage ranges from 3 to 4.15 V. The three NMC batteries are aged at accelerated calendar ageing condition illustrated in Table 4.3. The cells are stored in the thermostat at 35, 40 and 45 °C, while the SoC is set to 50% for each battery during the calendar ageing. The accelerated calendar ageing test lasted for 360 days, and a performance test was carried out every 30 days to measure the battery capacity at present. During the performance test, the ambient temperature is set to 25 °C and the sampling time is 1 s.

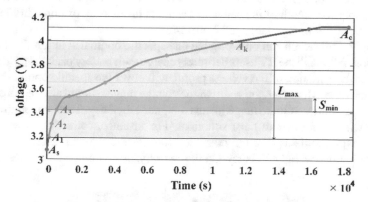

Fig. 4.13 Optimal signal voltage selection with grid search, reprinted from [53], with permission from Elsevier

Table 4.3 Accelerated calendar ageing condition

Temperature (°C)	35	40	45
SoC = 50%	Cell 1	Cell 2	Cell 3

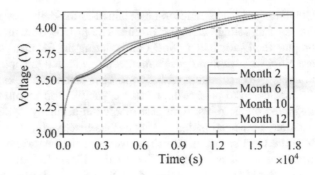

Fig. 4.14 The voltage curves of Cell 1 during the calendar ageing, reprinted from [53], with permission from Elsevier

The voltage profile of Cell 1 during the degradation test is shown in Fig. 4.14. The voltage measurement from Cell 2 and 3 shows a similar result. The voltage profiles gradually shift to the vertical axis because less energy can be stored in an aged cell.

Once grid search is used, one optimal voltage range can be found. We set the search step S_{min} to 0.1 V. Then, we find the optimal voltage ranges of the three NMC cells as shown in Fig. 4.15. According to these selected voltage ranges, the estimation results of the three cells are obtained as illustrated in Fig. 4.16.

The effectiveness of battery capacity estimation with one optimal voltage profile is proved by the results in Fig. 4.17. The estimation results are very close to the reference during the degradation procedure. The MSE is 2.5227×10^{-5} for Cell 1, 2.3441×10^{-5} for Cell 2 and 1.5151×10^{-5} for Cell 3. The optimal voltage range of Cell 1 in the first month of the calendar ageing is taken as an example in Fig. 4.17. The length of the voltage range is 12,669 s.

Notably, grid search can only provide one specific optimal solution for the SOH estimation. Although the EV users definitely charge their battery pack [54], it does not mean that the variation of the voltage profile will cover the specific optimal voltage range each time. Moreover, the width of the voltage ranges should also be considered in reality. A shorter voltage range means a higher efficiency. The two requirements conflict with each other in most situations. For instance, collecting measurement of the voltage profile as much as possible enhances the estimation accuracy, while deteriorating the overall efficiency. On the contrary, using limited data is easy for the

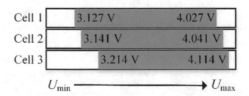

Fig. 4.15 Optimal voltage ranges of the three NMC cells, reprinted from [53], with permission from Elsevier

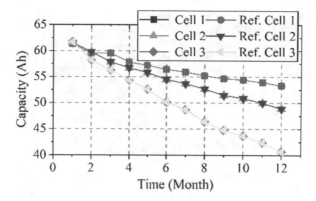

Fig. 4.16 Estimation results of the three cells, reprinted from [53], with permission from Elsevier

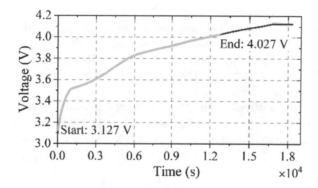

Fig. 4.17 Optimal voltage range of Cell 1, reprinted from [53], with permission from Elsevier

measurement, but the estimation accuracy may not be guaranteed. Hence, we need to solve a bi-objective optimization problem here for the best trade-off solutions. In order to conveniently obtain the voltage range in real applications, NSGA-II is used to choose two optimal voltage ranges considering both the length of the voltage profile and the accuracy of the capacity estimation. In addition, a series of non-dominated solutions from NSGA-II provide more freedom for BMS to estimate the battery SOH at various ageing stages.

The procedure of NSGA-II finding the two optimal voltage intervals are illustrated in Fig. 4.18. The voltage curve during the battery degradation is collected to form the original dataset. Afterwards, the initial populations are created by NSGA-II, which can reach multiple optimal solutions within one iteration. After evaluating the fitness of each individual, a fast non-dominated sorting algorithm is applied to assign the non-dominated level of each candidate solution. Additionally, the crowding distance is given to each individual. The new populations are selected from the best non-dominated set, and the solutions in the same non-dominated level are evaluated by the crowded comparison operator. Selection, crossover, and mutation are used to

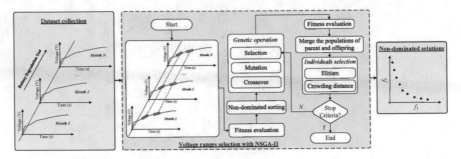

Fig. 4.18 Procedure of two voltage ranges selection with NSGA-II, reprinted from [53], with permission from Elsevier

generate the offspring from the current populations. Once the stop criteria are met, the partial charging voltages are found by NSGA-II.

In NSGA-II, each individual is encoded into a chromosome-like structure as shown in Fig. 4.19. The chromosome-like structure in Fig. 4.19a consists of four numbers U_{A1}, U_{A2}, U_{B1}, U_{B2}. From Fig. 4.19b, we know that U_{A1} and U_{A2} are the start and end points of the first voltage range, U_{B1} and U_{B2} are the corresponding points for the second voltage range.

The cost function of NSGA-II is well designed to evaluate the fitness of each individual. As the main purpose of the proposed method is the estimation accuracy, MSE of the estimation is used as one of the cost functions in Eq. (4.36).

$$f_1 = \frac{1}{n} \sum_{i=1}^{n} \left(Q_i - \hat{Q}_i \right)^2 \tag{4.36}$$

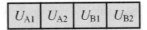

(a) The chromosome-like structure

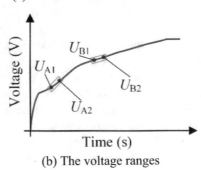

(b) The voltage ranges

Fig. 4.19 Individual representation, reprinted from [53], with permission from Elsevier

where Q_i is the reference capacity and $\hat{Q}_i$ is the estimated capacity, and n is number of the reference values during the degradation test. Thus, a smaller f_1 means a more accurate estimation. For the two voltage ranges condition, the battery capacity can be calculated as,

$$\widehat{Q}_i = \frac{\sum_{k=A_1}^{A_2} \eta \cdot I(k) \cdot T_s + \sum_{k=B_1}^{B_2} \eta \cdot I(k) \cdot T_s}{[\text{SoC}(A_2) - \text{SoC}(A_1)] + [\text{SoC}(B_2) - \text{SoC}(B_1)]} \tag{4.37}$$

The above equations mean that the SOC variations and the current integration of the two voltage ranges are accumulated, respectively. From the practical point of view, the length of the voltage ranges ($U_{A_1} \sim U_{A_2}$ and $U_{B_1} \sim U_{B_2}$) should be as small as possible for the purpose of conveniently obtaining the measurement from real applications. Thus, the cost function f_2 is defined as,

$$f_2 = \frac{1}{n} \sum_{i=1}^{n} \left(L_{U_{A1}-U_{A2}} + L_{U_{B1}-U_{B2}} \right) \tag{4.38}$$

where $L_{U_{A1}-U_{A2}}$ is the length between U_{A1} and U_{A2}, and $L_{U_{B1}-U_{B2}}$ is the length between U_{B1} and U_{B2}. Because the length of the voltage charging profile changes from week to week, the average length is chosen to calculate the f_2 during the battery degradation. Thus, a smaller f_2 is preferred because less voltage measurement is needed in this condition.

In practical, the two voltage ranges are time series measurement, and U_{A2} may not less than U_{A1} in a single range. Therefore, the constraints of the proposed method can be expressed as,

$$U_{A2} > U_{A1} \quad \text{and} \quad U_{B2} > U_{B1} \tag{4.39}$$

In this data science-based method, a special designed operator is applied to discard those illegal solutions in the selection operation. That is, a new solution will be selected only if the condition [Eq. (4.39)] is fulfilled. Otherwise, this solution has to be discarded, and the variation will be repeated until the new created solution is suitable for the constraints.

The value of f_1 and f_2 are in different ranges, and they are normalized between 0 and 1 for a better illustration, as shown in Fig. 4.20. All the solutions in the Pareto front of Fig. 4.20 are the optimal choice from a specific point of view. In Fig. 4.20, 50 non-dominated solutions form the Pareto front. Hence, various candidate solutions can be used to estimate battery SoH at different charging stages.

In this subsection, we only choose to show three typical solutions of Cell 1 in Fig. 4.21. The MSEs of each typical solution are listed in Table 4.4. The three typical solutions include two long voltage ranges (Solution A) and two short voltage ranges (Solution B and C). In Fig. 4.21, the two voltage ranges have some overlap for Solution A, while the two voltage ranges are separated and much shorter for

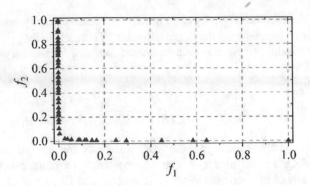

Fig. 4.20 Non-dominated solutions from NSGA-II, reprinted from [53], with permission from Elsevier

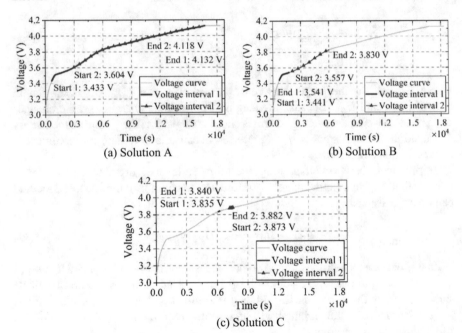

Fig. 4.21 Three typical solutions of Cell 1, reprinted from [53], with permission from Elsevier

Table 4.4 MSEs of the three typical solutions

Solution	A	B	C
MSE	1.75×10^{-5}	6.97×10^{-4}	8.96×10^{-2}

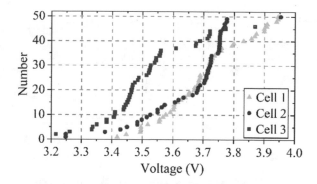

Fig. 4.22 Starting points of the non-dominated solutions of each cell, reprinted from [53], with permission from Elsevier

Solutions B and C. Compared with single voltage range, two voltage ranges provide more flexibility to estimate the SoH with partial charging profile.

The non-dominated solutions in different cells have some similarities as shown in Fig. 4.22. We can find that the starting points of the three cells are quite close to each other. For the starting points of one cell, there are always a starting point nearby for the other two cells. This indicts the generalization of the solutions from Cell 1 to Cells 2 and 3.

In order to further verify the generalization of the proposed method, the optimal voltage ranges from Cell 1 are directly applied to estimate the capacity of Cells 2 and 3. The three typical solutions of Cell 1 in Fig. 4.21 are validated by Cells 2 and 3. We can see the three typical solutions from Cell 1 also receive accuracy capacity estimation of Cells 2 and 3 in Fig. 4.23, which proves the generalization of the proposed method.

In the previous validation of this section, the SoC in the calculation comes from the Coulomb counting method with a known initial value. The reason is that the NMC batteries are always fully charged or discharged in our life time test. However, an initial SoC is hardly to be known in the real applications. The most popular estimation method in this area is the model-based SoC estimation methods, which can only provide less than $\pm 2\%$ error band. A $\pm 5\%$ error band is added to the SoC for verifying the accuracy of the proposed method. The denominator of Eq. (4.35) is the subtraction of $SoC(A_1)$ and $SoC(A_2)$, and the maximum and minimum of ΔSoC is expressed as:

$$\begin{cases} \Delta SoC_{min} = SoC_{-5\%}(A_2) - SoC_{+5\%}(A_1) \\ \Delta SoC_{max} = SoC_{+5\%}(A_2) - SoC_{-5\%}(A_1) \end{cases} \tag{4.40}$$

where $SoC_{-5\%}(\cdot)$ is the SOC including -5% error, and $SoC_{+5\%}(\cdot)$ is the SoC with $+5\%$ error.

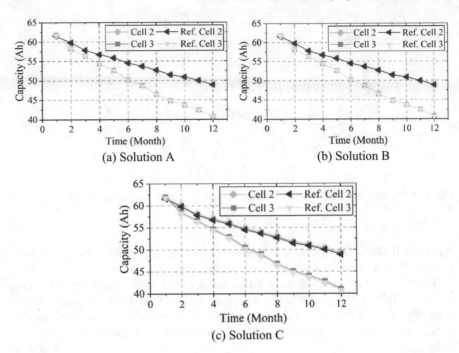

Fig. 4.23 Validation of three typical solutions from Cell 1 on Cells 2 and 3, reprinted from [53], with permission from Elsevier

In order to verify the effect of SoC estimation error on the performance of the proposed method, the voltage range [3.127 V, 4.027 V] is applied to estimate the capacity of Cell 1. The estimation results in Fig. 4.24 give the error band of the proposed method. Although ±5% SoC estimation error is added, the maximum absolute error of the estimation results is 3.2539 Ah, which is 5.16% of the battery

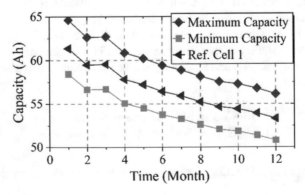

Fig. 4.24 Effect of SOC estimation error on the proposed method, reprinted from [53], with permission from Elsevier

nominal capacity as shown in Fig. 4.24. Thus, we can ensure the capacity estimation error of the proposed method is still in a small limited range when large SoC estimation error exists.

4.2.4 Joint State Estimation

Currently, there are a great deal of works focus on battery single state estimation in the literature, whereas the researches of joint estimation (co-estimations of at least two states) of battery multi-states are limited. It should be known that during operations, battery states would be coupled and interact with each other. Estimating just one state without considering others would cause only relatively satisfactory results under a certain constraint. In this context, for better operation management of batteries, the joint state estimation of battery considering the effects of different internal states is urgently required.

4.2.4.1 Definition of Battery Joint State Estimation

Figure 4.25 illustrates the relations of several strong-coupled battery critical states. Specifically, due to the fast-variations of battery electric dynamics, battery SoC and SoP would rapidly change with a short-term timescale. According to the battery physics structure and heat transfer nature, battery macroscopic states such as state

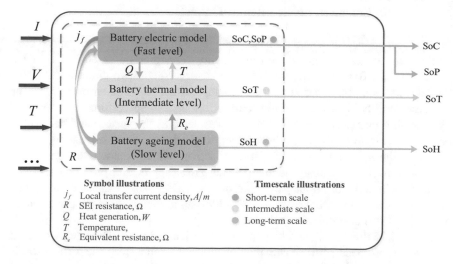

Fig. 4.25 Several battery key internal states with different timescale (here I, V, and T_f represent battery terminal current, voltage and surface temperature, respectively), reprinted from [32], with permission from Elsevier

of temperature (SoT) would change with an intermediate timescale [55]. For battery SoH, as it is manifested by several slow-variation factors such as internal impedance or resistance increase and capacity degradation, this state would change slowly with a long-term timescale during battery operations.

To date, just a few existing data science researches focus on double-states co-estimations of battery. Among these researches, the joint estimation of both battery SoC and SoH plays a dominant position. This is primarily caused by the fact that updating battery SoH information (capacity or resistance) periodically is crucial for enhancing the estimation performance of battery SoC. Based upon the equivalent circuit models or electrochemical models, various data science observers including the Kalman filter (KF) [46], adaptive filter [56], and their variants, such as the extended KF (EKF) [57], dual-fractional-order extended KF (DEKF) [46], have been designed to effectively co-estimating battery SoC and SoH simultaneously. Besides, apart from joint-estimating battery SoC and SoH, limited data science research has been also done to estimate other battery double-states, such as the co-estimations of battery SoC and SoP [58], SoC and SoT [59]. Furthermore, compared with only one battery state estimation, larger computational burden is generally required for joint state estimation applications. In this context, to widen battery joint state estimations, state-of-the-art data science solutions such as the fractional order calculus [46] and multi-timescale estimators [60] that could enhance co-estimation accuracy and provide a satisfactory computational effort are becoming a promising research direction.

4.2.4.2 Case Study: Battery SoC and SoH Co-estimation with Enhanced Electrochemical Model

In this subsection, a data science case study through developing an enhanced electrochemical model to achieve the high-fidelity co-estimation of SoC and SoH is presented [61]. To be specific, the full-order battery Pseudo-two-dimensional (P2D) model is first simplified based on the Padé approximation while ensuring precision and observability. Next, the feasibility and performance of SoC estimator are revealed by accessing unmeasurable physical variables, such as the surface and bulk solid-phase concentration. To well reflect battery degradation, three key ageing factors including the loss of lithium ions, loss of active volume fraction, and resistance increment are simultaneously identified, leading to an appreciable precision improvement of SoC estimation online particular for aged cells. Finally, extensive verification experiments are carried out over the cell's lifespan to demonstrate the performance of this SoC/SoH co-estimation scheme.

Figure 4.26 illustrates the schematic of the typical battery P2D model, where the Li-ions are assumed to diffuse with the directions of x and r. Here, the particle would be uniformly distributed with a radius of R_s. In general, four governing equations including the conservation of Li^+ and charge in both solid as well as electrolyte phases are adopted to formulate a battery electrochemical model. The diffusion of Li^+ within a single particle is generally captured by the Fick's law as [62]:

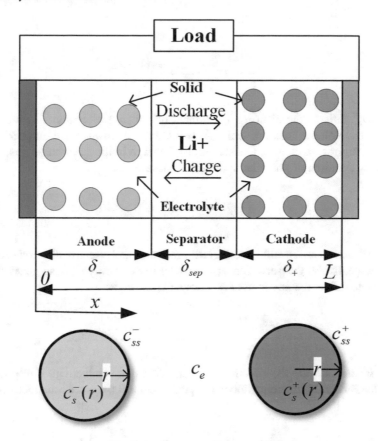

Fig. 4.26 Schematic of the typical battery P2D model, reprinted from [61], with permission from IEEE

$$\frac{\partial c_s}{\partial t} = \frac{D_s}{r^2} \frac{\partial}{\partial r} \left(r^2 \frac{\partial c_s}{\partial r} \right) \tag{4.41}$$

where D_s is a coefficient to reflect solid diffusion, and c_s represents the concentration of Li$^+$ within solid phase. c_{ss} means the concentration of particle surface at $r = R_s$. $c_{ss}(t) = c_s(R_s, t)$. The bulk concentration $\overline{c}_s^{\pm}$ in the anode/cathode can be obtained by:

$$\frac{\partial \overline{c}_s}{\partial t} = \frac{I}{F A_{\text{cell}} \delta \varepsilon_s} \tag{4.42}$$

where I reflects input current. F and A_{cell} are Faraday constant and electrode surface area, respectively. δ stands for electrode thickness, while ε_s means the active materials' volume fraction. The lithium concentration c_e of electrolyte can be expressed by:

$$\varepsilon_e \frac{\partial c_e}{\partial t} = D_e^{\text{eff}} \frac{\partial^2 c_e}{\partial x^2} + \frac{a_s\left(1 - t_+^0\right)}{F} j \tag{4.43}$$

where D_e^{eff} reflects the diffusion coefficient of effective electrolyte. ε_e means the electrolyte's volume fraction, while t_+^0 stands for the transference number of Li$^+$. j is the lithium flux density. Conservation of charge in the solid phase generates a governing equation of potential in the solid phase ϕ_s as:

$$\sigma^{\text{eff}} \frac{\partial^2 \phi_s}{\partial x^2} - a_s j = 0 \tag{4.44}$$

where σ^{eff} and a_s are specific interfacial surface area and effective electrode conductivity, respectively. Conservation of charge in the electrolyte phase generates the equation to reflect the potential of electrolyte phase ϕ_e as:

$$k^{\text{eff}} \frac{\partial^2 \phi_e}{\partial x^2} + k_d^{\text{eff}} \frac{\partial^2 \ln c_e}{\partial x^2} + a_s j = 0 \tag{4.45}$$

where k^{eff} and k_d^{eff} are the effective ionic and diffusion conductivities, respectively. Here, the Butler–Volmer equation is adopted to control the electrochemical kinetics as:

$$j = i_0 \left(\exp\left(\frac{\alpha_a F}{RT}\eta\right) - \exp\left(-\frac{\alpha_c F}{RT}\eta\right) \right) \tag{4.46}$$

where i_0 is the exchange current density. α_a and α_c are coefficients to reflect anode and cathode charge transfer, respectively. R and T are universal gas constant and temperature, respectively. Overpotential η is the extra force needed to overcome surface reaction by:

$$\eta_p(t) = \phi_s(L, t) - \phi_e(L, t) - U_p(c_{ss}^+)$$
$$\eta_n(t) = \phi_s(0, t) - \phi_e(0, t) - U_n(c_{ss}^-) \tag{4.47}$$

where U_p is cathode open-circuit potential, and U_n is anode open-circuit potential. Here the cell terminal voltage could be described by:

$$V(t) = \phi_s(L, t) - \phi_s(0, t) - R_f I \tag{4.48}$$

where R_f stands for the summation of solid electrolyte interface (SEI) resistance and ohmic resistance [63].

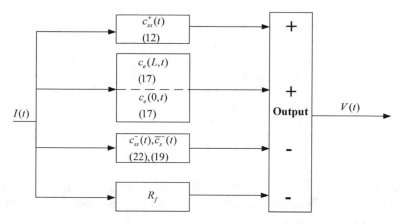

Fig. 4.27 Diagram of simplified battery electrochemical model, reprinted from [61], with permission from IEEE

The final expression of terminal voltage could be obtained by substituting Eq. (4.47) into Eq. (4.48):

$$V(t) = (U_p(c_{ss}^+) + \eta_p(t) + \phi_e(L, t))$$
$$- (U_n(c_{ss}^-) + \eta_n(t) + \phi_e(0, t)) - R_f I(t) \qquad (4.49)$$

Based upon the above discussion, Fig. 4.27 illustrates the block diagram of battery reduced-order electrochemical model as:

To simplify electrochemical model, Eq. (4.41) could be further modified after taking Laplace transform as:

$$\frac{C_{ss}(R_s, s)}{J(s)} = \frac{R_s}{a_s F D_s} \frac{\tanh\left(\sqrt{\frac{s}{D_s}} R_s\right)}{\tanh\left(\sqrt{\frac{s}{D_s}} R_s\right) - \sqrt{\frac{s}{D_s}} R_s} \qquad (4.50)$$

where a_s, F, and D_s are the specific interfacial area, Faraday constant, and diffusion coefficient of solid-phase Li$^+$, respectively.

Then a third-order Padé approximation [64] is adopted to convert Eq. (4.50) into a polynomial transfer function as:

$$\frac{C_{ss}(s)}{J(s)} = \pm \frac{\frac{3}{R_s} + \frac{4R_s}{11D_s} s + \frac{R_s^3}{165D_s^2} s^2}{a_s F \left(s + \frac{3R_s^2}{55D_s} s^2 + \frac{R_s^4}{3465D_s^2} s^3\right)} \qquad (4.51)$$

Equation (4.51) can be further transformed into a state-space equation with the controller canonical form as:

$$\dot{x}_i = A_i x_i + B_i u$$

$$A_i = \begin{bmatrix} 0 & 1 & 0 \\ 0 & 0 & 1 \\ 0 & -\dfrac{3465 D_s^2}{R_s^4} & -\dfrac{2079}{11 R_s^2} \end{bmatrix}, \quad B_i = \begin{bmatrix} 0 \\ 0 \\ \pm\dfrac{3465 D_s^2}{a_s F R_s^4} \end{bmatrix}$$

$$c_{ss} = \begin{bmatrix} \dfrac{3}{R_s} & \dfrac{4 R_s}{11 D_s} & \dfrac{R_s^3}{165 D_s^2} \end{bmatrix} x_i$$

$$\bar{c}_s = \begin{bmatrix} \dfrac{3}{R_s} & \dfrac{9 R_s}{55 D_s} & \dfrac{R_s^3}{1155 D_s^2} \end{bmatrix} x_i \tag{4.52}$$

where $i = \{p, n\}$, $x_i = [x_1\ x_2\ x_3]^T$, u is the input current. Here, x_1, x_2, and x_3 are utilized to describe the surface c_{ss} and bulk $\bar{c}_s$ lithium concentration in the electrodes without physical meanings.

According to the Taylor expansion, the Butler–Volmer equation could be linearized by:

$$\eta(s) = \frac{RT}{F i_0(\alpha_a + \alpha_c)} J(s) \tag{4.53}$$

where α_a is the symmetric anodic reaction charge transfer coefficient, and α_c the symmetric cathodic one. i_0 stands for the exchange current density that is correlated with ion concentrations by: $i_0 = k(C_e)^{\alpha_a}(C_{s,\max} - C_{ss})^{\alpha_a}(C_{s,\max} - C_{ss})^{\alpha_c}$.

Here, the intercalation current density of electrode is proportional to battery current as:

$$J(s) = \frac{I}{A_{\text{cell}} F \delta a_s} \tag{4.54}$$

After adopting the analytical solution with first-order Padé approximation [45], electrolyte potential difference from Eq. (4.45) could be derived by:

$$\frac{C_e(L, s)}{J_p(s)} = \frac{1}{b_{1,p} s + b_{2,p}} \tag{4.55}$$

$$\frac{C_e(0, s)}{J_n(s)} = \frac{1}{b_{1,n} s + b_{2,n}} \tag{4.56}$$

Here, $b_{1,p}$, $b_{2,p}$, $b_{1,n}$, and $b_{2,n}$ are all constant parameters. Equations (4.55) and (4.56) could be expressed by state-space form as:

$$\dot{x}_e = A x_e + B u$$

$$A = \begin{bmatrix} -\dfrac{b_{2,p}}{b_{1,p}} & 0 \\ 0 & -\dfrac{b_{2,n}}{b_{1,n}} \end{bmatrix}, \quad B = \begin{bmatrix} \dfrac{1}{b_{1,p}} \\ \dfrac{1}{b_{2,n}} \end{bmatrix}$$

$$\begin{bmatrix} c_{e,L} \\ c_{e,0} \end{bmatrix} = \begin{bmatrix} 1 & 1 \end{bmatrix} x_e \tag{4.57}$$

where u stands for battery current, x_e is a vector with one-dimension to represent the electrolyte concentration at $x = 0/L$.

Next, the derived data science observer for the co-estimation of battery SoC and SoH is proposed in detail. Specifically, after the above simplification of P2D model, battery terminal voltage from Eq. (4.49) can be described by:

$$\begin{aligned} V(t) = {} & U_p(c_{ss}^+) - U_n(c_{ss}^-) \\ & + \eta_p(i_{0,p}(c_{ss}^+, c_{e,p})) - \eta_n(i_{0,p}(c_{ss}^-, c_{e,n})) \\ & + \phi_e(c_{e,L}) - \phi_e(c_{e,0}) - R_f I \end{aligned} \tag{4.58}$$

To decrease the condition number for battery SoC estimator, one effective data science solution through estimating the lithium concentration of negative electrode with the open-loop simulations of positive electrode as well as liquid phase is adopted. In this context, the positive electrode and liquid potentials subtracted from battery terminal voltage could be utilized as the feedback of anode observer. According to Eq. (4.52), the model formulation of SoC estimation part is:

$$\begin{aligned} \dot{x}_n = f(x_n, u) = A x_n + B u \\ A = \begin{bmatrix} 0 & 1 & 0 \\ 0 & 0 & 1 \\ 0 & -\dfrac{3465 D_{s,n}^2}{R_{s,n}^4} & -\dfrac{2079}{11 R_{s,n}^2} \end{bmatrix}, \quad B = \begin{bmatrix} 0 \\ 0 \\ -\dfrac{3465 D_{s,n}^2}{a_{s,n} F R_{s,n}^4} \end{bmatrix} \\ g(x_n, u) = \phi_s^- = U_n + \eta_n + R_f I \end{aligned} \tag{4.59}$$

where $x_n = [x_1 \; x_2 \; x_3]^T$, u represents the input current.

For battery SoH estimation part, the recyclable lithium loss caused by the side reaction of anode would lead to the shift of $\theta_{100\%,n}$ (upper voltage limits) and $\theta_{0\%,n}$ (lower voltage limits), respectively. In this context, lithium ions loss could be determined through estimating the concentration of normalized anode bulk with the fully charging or discharging state. Here, the analogical open-loop framework would be adopted to facilitate anode observer. With regard to active material loss and internal resistance increase, the estimation of these two ageing factors becomes the monitoring of $\varepsilon_{s,\text{neg}}$ and R_f within the model. Therefore, $\varepsilon_{s,\text{neg}}$ and R_f are treated as another states within anode observer in the designed joint estimation framework. In the light of this, equations to describe state dynamics of anode are expressed as:

$$\dot{x}_\theta = f(x_\theta, u) = A x_\theta + B u$$

$$A = \begin{bmatrix} 0 & 0 & 0 \\ 0 & 0 & 0 \\ 0 & 0 & 0 \end{bmatrix}, \quad B = \begin{bmatrix} \frac{-1}{F A_{\text{cell}} \delta_{-\varepsilon_{s,n}}} \\ 0 \\ 0 \end{bmatrix}$$

$$g(x_\theta, u) = \phi_s^- = U_n + \eta_n + R_f I \tag{4.60}$$

where $x_\theta = [\theta_1 \ \theta_2 \ \theta_3]^T$, u represents battery current. Here, θ_1 would determine bulk concentration of graphite electrode. The time derivatives of anode's bulk concentration have been shown in Eq. (4.60). θ_2 and θ_3 represent the active material volume of anode $\varepsilon_{s,n}$ and interior resistance R_f, respectively. The time derivatives of $\varepsilon_{s,n}$ and R_f are set to zero. Afterwards, the observability of derived SoC and SoH co-estimator could be described by the *Lie derivatives*. Here, the simplified battery model equations is reformulated by:

$$\dot{x}_{n/\theta} = f(x_{n/\theta}, u) + w$$
$$y = g(x_{n/\theta}, u) + v \tag{4.61}$$

where x and u represent state and input, respectively. w is system noise while v is measurement noise. The $N - 1$ order *Lie derivatives* of g is expressed by:

$$L_f^0(g) = g(x_{n/\theta}, u)$$
$$\vdots$$
$$L_f^{N-1}(g) = g^{N-1}(x_{n/\theta}, u, \dot{u}, \ldots, u^{(N-1)}) \tag{4.62}$$

According to the Jacobian of *Lie derivatives* set [23], the observability matrix could by expressed as:

$$\Theta = \begin{bmatrix} \frac{\partial L_f^0(g)}{\partial x_{n/\theta,1}} & \cdots & \frac{\partial L_f^0(g)}{\partial x_{n/\theta,N}} \\ \vdots & \ddots & \vdots \\ \frac{\partial L_f^{N-1}(g)}{\partial x_{n/\theta,1}} & \cdots & \frac{\partial L_f^{N-1}(g)}{\partial x_{n/\theta,N}} \end{bmatrix} \tag{4.63}$$

where $x_{n/\theta,N}$ represents the Nth element of $x_{n/\theta}$.

To simultaneously estimate both battery states and parameters of the simplified electrochemical model, a dual extended Kalman filter (DEKF) is adopted. To be specific, SoC estimation under a fast timescale is realized by one filter based on the known ageing parameters, while SoH estimation under slow timescale is realized by using another filter to identify related ageing parameters online. Figure 4.28 details the whole scheme for battery SoC and SoH co-estimation, where the corresponding parameters are described in Table 4.5. Here, Q and R are the covariance matrixes of process and sensor noises. All four ageing parameters $\theta_{100\%,n}$, $\theta_{0\%,n}$, $\varepsilon_{s,n}$ and R_f in

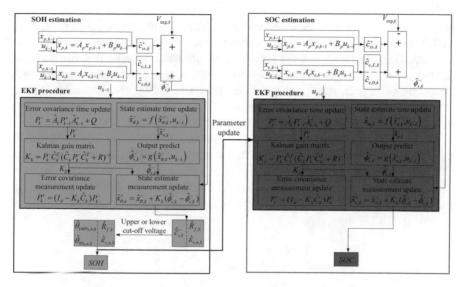

Fig. 4.28 Schematic of joint estimation of battery SoC and SoH, reprinted from [61], with permission from IEEE

Table 4.5 EKF parameters and procedure, reprinted from [61], with permission from IEEE

	EKF for SoC estimation	EKF for SoH estimation
$\widehat{A}_k$	$\frac{\partial f(x_{n,k}, u_k)}{\partial x_{n,k}}\big\|x_{n,k} = \hat{x}^+_{n,k}$	$\frac{\partial f(x_{\theta,k}, u_k)}{\partial x_{\theta,k}}\big\|x_{\theta,k} = \hat{x}^+_{\theta,k}$
$\widehat{C}_k$	$\frac{\partial g(x_{n,k}, u_k)}{\partial x_{n,k}}\big\|x_{n,k} = \hat{x}^-_{n,k}$	$\frac{\partial g(x_{\theta,k}, u_k)}{\partial x_{\theta,k}}\big\|x_{\theta,k} = \hat{x}^-_{\theta,k}$
Q	$10^{-9} \times \mathrm{diag}(4, 2, 27)$	$10^{-16} \times \mathrm{diag}(4, 2800, 9000)$
R	4×10^{-3}	6×10^{-2}

Initialization for $k = 0$

$$\hat{x}^+_{n/\theta,0} = E[x_{n/\theta,0}]$$

$$P_0^+ = E[\left(x_{n/\theta,0} - \hat{x}^+_{n/\theta,0}\right)\left(x_{n/\theta,0} - \hat{x}^+_{n/\theta,0}\right)^T]$$

Iteration for $k = 1, 2, \ldots$

State-prediction time update: $\hat{x}^-_{n/\theta,k} = f\left(\hat{x}^+_{n/\theta,k}, u_{k-1}\right)$

Error-covariance time update: $P_k^- = \widehat{A}_k P_{k-1}^+ \widehat{A}_{k-1}^+ + Q$

Output estimate: $\hat{y}_k^- = g(\hat{x}^-_{n/\theta,k}, u_{k-1})$

Estimator gain matrix: $K_k = P_k^- \hat{C}_k^T (\widehat{C}_k P_k^- \widehat{C}_k^T + R)^{-1}$

State-estimate measurement update: $\hat{x}^+_{n/\theta,k} = \hat{x}^-_{n/\theta,k} + K_k(y_k - \hat{y}_k^-)$

Error-covariance measurement update: $P_k^+ = (I_d - K_k \widehat{C}_k) P_k^-$

battery SoC estimation block would be periodically updated from related predictions of SoH estimation block, further benefitting the accuracy of SoC estimator over time.

Next, the effectiveness of the proposed co-estimation scheme is verified against the experimental results of battery cycling tests. A series of characterization tests, including a dynamic stress test (DST) cycle and an Urban Dynamometer Driving Schedule (UDDS) cycle, are performed on the NCM/graphite18650 batteries every two weeks. The cycling test schedule is depicted in Fig. 4.29.

Before implementing the SoC/SoH co-estimation algorithm, it is necessary to validate the accuracy of electrochemical model. Figure 4.30 illustrates the voltage responses of the developed model with both static load (1C CCCV charging, CC discharging) and dynamic load (UDDS, DST cycles with 3C maximum current). The predicted voltage shows good agreement with the measured voltage. The simplified

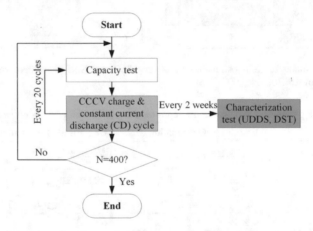

Fig. 4.29 Cyclic test process to generate battery experimental data, reprinted from [61], with permission from IEEE

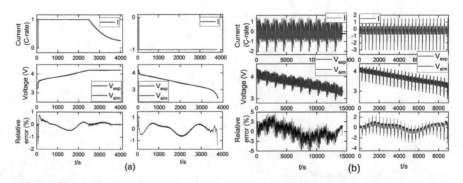

Fig. 4.30 Experimental validation of enhanced electrochemical model: **a** 1C CCCV charging and CC discharging, **b** UDDS cycles with 2C maximum current, and DST cycles with 3C maximum current, reprinted from [61], with permission from IEEE

model achieves a voltage RMSE of {10.52, 11.85, 23.25, 21.34 mV} for these cases of CCCV charge, CC discharge, UDDS cycles, and DST cycles, respectively. The results indicate that the model parameters are initialized properly.

To verify the effectiveness of proposed data science method for SoH estimation, a single CC discharging and CCCV charging cycle is adopted. Here, the state and parameter estimates are randomly initialized as: $\hat{\theta}_n(0) = 0.7\theta_n^*(0)$, $\hat{\varepsilon}_{s,n}(0) = 0.5\varepsilon_s^*$, and $\widehat{R}_f(0) = 2R_f^*$. After inputting the measured voltage and current, the evolutions of both state and parameter estimates are illustrated in Fig. 4.31. It can be seen that although there exists large initial errors, $\hat{\theta}_n$, $\hat{\varepsilon}_{s,n}$, and $\widehat{R}_f$ could converge to their nominal values within 3000 s of discharging process.

Figure 4.32a–d illustrates the capacity estimation results of four cells with different ageing levels. For fresh cell in Fig. 4.32a, the capacity estimates with CD-CCCV, UDDS, and DST tests all converge to the measured capacity gradually even these estimates are initialized with incorrect values, indicating the effectiveness of the observer. Similar estimation performance can be achieved at ageing levels 2, 3, and 4, as shown in Fig. 4.32b–d. Figure 4.33a and b shows the capacity estimation perfor-

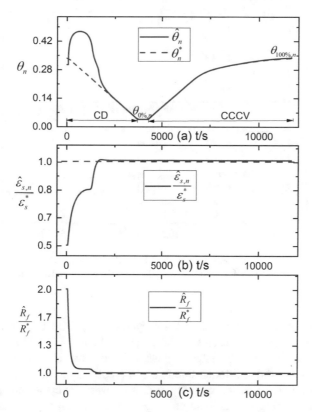

Fig. 4.31 Evolution of estimated parameters with CC discharging-CCCV charging cycle: **a** $\theta_{100\%,n}$, $\theta_{0\%,n}$, **b** $\varepsilon_{s,n}$, and **c** R_f, reprinted from [61], with permission from IEEE

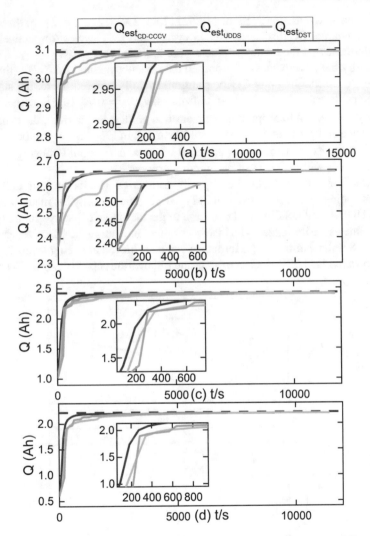

Fig. 4.32 Battery SoH estimation results with CD-CCCV, UDDS, DST tests at different ageing levels, **a** 100% SoH, **b** 86% SoH, **c** 78% SoH, and **d** 71% SoH, reprinted from [61], with permission from IEEE

mance of the derived estimator across 400 cycles under CC-CCCV scenario. Here, the capacity is predicted at each CC-CCCV cycle. Figure 4.33a plots the values of estimated capacities with blue plus symbols. The red dot symbols represent capacity measurements and the red dotted line is the interpolated curve using the measured data. It can be noted that the capacity estimates over 400 cycles could follow the red curve closely. Figure 4.33b summarizes the related capacity estimation error. Here, the mean error is 0.0241 Ah, which is within 1% of the cell nominal capacity.

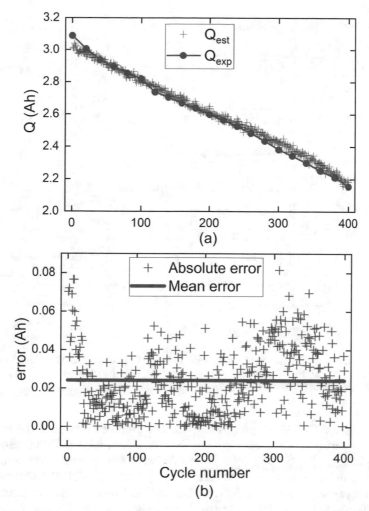

Fig. 4.33 Battery capacity estimation results: **a** results against reference values, **b** estimation error, reprinted from [61], with permission from IEEE

The statistical results reveal that the SoH estimations with the presented algorithm achieve agreeable precision over the cell lifespan.

Apart from battery SoH estimation, another key task is to estimate the variations of model parameters that relate to the dominant ageing mechanisms within a battery (loss of recyclable ions, loss of active materials, and resistance increase), thereby ensuring an accurate SoC estimate overtime. To examine the robustness of the proposed co-estimation scheme against battery ageing, the cells at four various ageing levels with the capacity of 3.088, 2.658, 2.406, and 2.199 Ah are examined. Both UDDS and DST cycles in the characterization tests are carried out to simulate the operating load profiles of dynamic EV applications. The cyclic data with CC

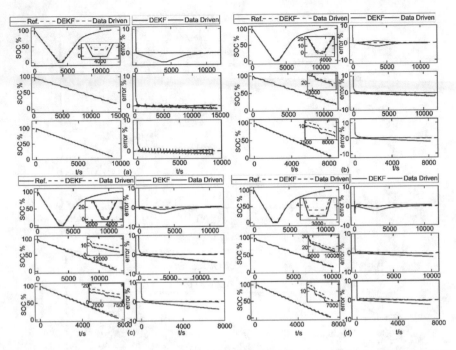

Fig. 4.34 Battery SoC estimation results with CD-CCCV (upper), UDDS (middle) and DST (bottom) cycles, **a** 100% SoH, **b** 86% SoH, **c** 78% SoH, and **d** 71% SoH, reprinted from [61], with permission from IEEE

discharging and CCCV charging is utilized. The proposed dual EKF (DEKF) SoC estimation performance is compared to the data-driven joint SoH/SoC estimation for CC-CCCV, UDDS, DST tests at four different ageing levels. The framework of the data-driven joint estimation is inspired by Refs. [65–67]. The obtained SoC estimation results under CC-CCCV, UDDS, and DST profiles with electrochemical model-based DEKF and data-driven method are shown in Fig. 4.34a–d. All the SoC estimations with DEKF are initialized with an error of 10%. For the fresh cell in Fig. 4.34a, although large initial errors of 10% are imposed, the online estimation of SoC can still quickly converge and show good agreement with the referenced SoC. According to Fig. 4.34b, when the battery parameters decay to 2.658 Ah (71% SoH), the DEKF SoC estimation performance outperforms the data-driven method. For DEKF SoC estimation, the estimation error with CC-CCCV test drops below 1% after 98 s and then converges towards less than 0.44% in a steady state. It can be noted that the SoC estimation results with UDDS and DST tests demonstrate fast convergence and high precision as well. However, the SoC estimation errors of data-driven SoC estimation are unable to converge to the true values with battery degradation. The maximum absolute errors reach 1.95%, 1.73%, and 2.28% with CC-CCCV, UDDS and DST tests, respectively. For cells with 2.406 and 2.199 Ah, similar trends can be observed as well. The data-driven SoC estimation steadily drifts away from its

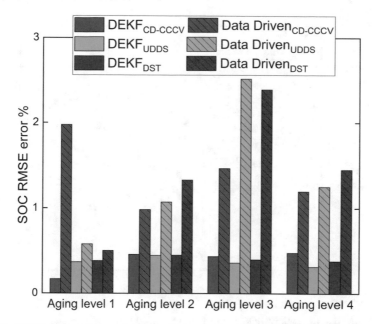

Fig. 4.35 RMSE errors of battery SoC estimation, reprinted from [61], with permission from IEEE

initial values. This leads to large errors beyond the accurate starting value of 100%. In contrast, the DEKF SoC estimation quickly converges to the reference and steadily finds its way back to a close neighbourhood of the reference value.

Figure 4.35 illustrates the RMSEs of SoC estimation at these ageing levels. Even for cells at the most degraded level 4, the corresponding RMSEs of SoC estimation with the proposed method are all within 0.48%. However, the data-driven SoC estimation results increase over 1.20%, indicating that the robustness of the data-driven SoC estimator should be strengthened by considering measurement noise and error compensation for aged cells.

Figure 4.36 shows the RMSEs of DEKF SoC estimation across 400 cycles under CC-CCCV scenario. The estimated $\theta_{0\%,n}$, $\theta_{100\%,n}$, $\varepsilon_{s,n}$ and R_f from the SoH observer are used to update the SoC estimator after each CC-CCCV cycle. The SoC estimate RMSE over a given CD-CCCV cycle almost remains constant near the averaged RMSEs throughout the life of the cell. The maximum SoC RMSE is less than 0.7%. This implies that the model is updated correctly with the estimates of $\theta_{0\%,n}$, $\theta_{100\%,n}$, $\varepsilon_{s,n}$ and R_f. It is clear that the co-estimation of SoC/SoH could not only provide electrochemical-mechanism enhanced SoH prediction at relatively slow timescale but also improve its real-time resilience to disturbance caused by battery degradation. Therefore, the SoC estimation performance is well sustained over the cell's lifespan.

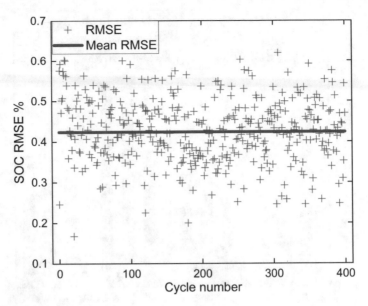

Fig. 4.36 RMSEs for battery SOC estimation versus 400 cycles, reprinted from [61], with permission from IEEE

4.3   Summary

This chapter mainly focuses on the data science-based battery operation modelling and state estimation, two basic parts for battery operation management. Specifically, three typical types of battery operation models including battery electrical model, battery thermal model, and battery coupled model are first described. Then, the fundamentals of battery SoC, SoP, SoH, and joint states estimations are introduced. The advantages and limitations of each mainstream type of state estimation method are compared and discussed. For SoC estimation, a data science-based case study using RLS and EKF is introduced. Then based upon the estimated SoC information, another data science-based case study of using the multi-constrained dynamic method to estimate battery SoP is also given. For SoH estimation, after using NSGA-II to select the optimal multiple voltage ranges, a data science-based case study is introduced to estimate SoH based on the optimized partial charging voltage profiles. After that, a data science-based case study through developing an enhanced electro-chemical model to achieve high-fidelity co-estimation of SoC and SoH is presented. All these case studies could give reasonable and effective estimation results, while co-estimation is able to present better performance. These results indicate that battery states are coupled and interact with each other during operations. Satisfactory battery operation modelling and state estimations can be achieved with suitable data science solutions.

References

1. Liu K, Li K, Peng Q, Zhang C (2019) A brief review on key technologies in the battery management system of electric vehicles. Front Mech Eng 14(1):47–64
2. Rahman MA, Anwar S, Izadian A (2016) Electrochemical model parameter identification of a lithium-ion battery using particle swarm optimization method. J Power Sources 307:86–97
3. Sung W, Shin CB (2015) Electrochemical model of a lithium-ion battery implemented into an automotive battery management system. Comput Chem Eng 76:87–97
4. Han X, Ouyang M, Lu L, Li J (2015) Simplification of physics-based electrochemical model for lithium ion battery on electric vehicle. Part II: Pseudo-two-dimensional model simplification and state of charge estimation. J Power Sources 278:814–825
5. Zou C, Manzie C, Nešić D (2015) A framework for simplification of PDE-based lithium-ion battery models. IEEE Trans Control Syst Technol 24(5):1594–1609
6. Zhang L, Wang Z, Hu X, Sun F, Dorrell DG (2015) A comparative study of equivalent circuit models of ultracapacitors for electric vehicles. J Power Sources 274:899–906
7. Nejad S, Gladwin D, Stone D (2016) A systematic review of lumped-parameter equivalent circuit models for real-time estimation of lithium-ion battery states. J Power Sources 316:183–196
8. Gong X, Xiong R, Mi CC (2015) A data-driven bias-correction-method-based lithium-ion battery modeling approach for electric vehicle applications. IEEE Trans Ind Appl 52(2):1759–1765
9. Wang Q-K, He Y-J, Shen J-N, Ma Z-F, Zhong G-B (2017) A unified modeling framework for lithium-ion batteries: an artificial neural network based thermal coupled equivalent circuit model approach. Energy 138:118–132
10. Deng Z, Yang L, Cai Y, Deng H, Sun L (2016) Online available capacity prediction and state of charge estimation based on advanced data-driven algorithms for lithium iron phosphate battery. Energy 112:469–480
11. Sbarufatti C, Corbetta M, Giglio M, Cadini F (2017) Adaptive prognosis of lithium-ion batteries based on the combination of particle filters and radial basis function neural networks. J Power Sources 344:128–140
12. Li Y, Chattopadhyay P, Xiong S, Ray A, Rahn CD (2016) Dynamic data-driven and model-based recursive analysis for estimation of battery state-of-charge. Appl Energy 184:266–275
13. Xie Y, Zheng J, Hu X, Lin X, Liu K, Sun J, Zhang Y, Dan D, Xi D, Feng F (2020) An improved resistance-based thermal model for prismatic lithium-ion battery charging. Appl Therm Eng 180:115794
14. Li W, Xie Y, Liu K, Yang R, Chen B, Zhang Y (in press) An enhanced thermal model with virtual resistance technique for pouch batteries at low temperature and high current rates. IEEE J Emerg Sel Topics Power Electron. https://doi.org/10.1109/JESTPE.2021.3127892
15. Shang Y, Liu K, Cui N, Zhang Q, Zhang C (2019) A sine-wave heating circuit for automotive battery self-heating at subzero temperatures. IEEE Trans Ind Inform 16(5):3355–3365
16. Raijmakers LH, Danilov DL, Van Lammeren JP, Lammers TJ, Bergveld HJ, Notten PH (2016) Non-zero intercept frequency: an accurate method to determine the integral temperature of Li-ion batteries. IEEE Trans Ind Electron 63(5):3168–3178
17. Lee K-T, Dai M-J, Chuang C-C (2017) Temperature-compensated model for lithium-ion polymer batteries with extended Kalman filter state-of-charge estimation for an implantable charger. IEEE Trans Ind Electron 65(1):589–596
18. Zhu C, Han J, Zhang H, Lu F, Liu K, Zhang X (in press) Modeling and control of an integrated self-heater for automotive batteries based on traction motor drive reconfiguration. IEEE J Emerg Sel Topics Power Electron. https://doi.org/10.1109/JESTPE.2021.3119599
19. Shang Y, Liu K, Cui N, Wang N, Li K, Zhang C (2019) A compact resonant switched-capacitor heater for lithium-ion battery self-heating at low temperatures. IEEE Trans Power Electron 35(7):7134–7144
20. Guo M, Kim G-H, White RE (2013) A three-dimensional multi-physics model for a Li-ion battery. J Power Sources 240:80–94

21. Jeon DH, Baek SM (2011) Thermal modeling of cylindrical lithium ion battery during discharge cycle. Energy Convers Manag 52(8–9):2973–2981
22. Jaguemont J, Omar N, Martel F, Van Den Bossche P, Van Mierlo J (2017) Streamline three-dimensional thermal model of a lithium titanate pouch cell battery in extreme temperature conditions with module simulation. J Power Sources 367:24–33
23. Shah K, Vishwakarma V, Jain A (2016) Measurement of multiscale thermal transport phenomena in Li-ion cells: a review. J Electrochem Energy Convers Storage 13(3)
24. Chen D, Jiang J, Li X, Wang Z, Zhang W (2016) Modeling of a pouch lithium ion battery using a distributed parameter equivalent circuit for internal non-uniformity analysis. Energies 9(11):865
25. Hu X, Asgari S, Yavuz I, Stanton S, Hsu CC, Shi ZY, Wang B, Chu HK, A transient reduced order model for battery thermal management based on singular value decomposition. In: Proceedings of IEEE energy conversion congress and exposition (ECCE), Pittsburgh, PA, 2014, pp 3971–3976
26. Lin X, Perez HE, Mohan S, Siegel JB, Stefanopoulou AG, Ding Y, Castanier MP (2014) A lumped-parameter electro-thermal model for cylindrical batteries. J Power Sources 257:1–11
27. Perez HE, Hu X, Dey S, Moura SJ (2017) Optimal charging of Li-ion batteries with coupled electro-thermal-aging dynamics. IEEE Trans Veh Technol 66(9):7761–7770
28. Dey S, Ayalew B (2017) Real-time estimation of lithium-ion concentration in both electrodes of a lithium-ion battery cell utilizing electrochemical–thermal coupling. J Dyn Syst Meas Control 139(3)
29. Goutam S, Nikolian A, Jaguemont J, Smekens J, Omar N, Bossche PVD, Van Mierlo J (2017) Three-dimensional electro-thermal model of Li-ion pouch cell: analysis and comparison of cell design factors and model assumptions. Appl Therm Eng 126:796–808
30. Jiang J, Ruan H, Sun B, Zhang W, Gao W, Zhang L (2016) A reduced low-temperature electro-thermal coupled model for lithium-ion batteries. Appl Energy 177:804–816
31. Basu S, Hariharan KS, Kolake SM, Song T, Sohn DK, Yeo T (2016) Coupled electrochemical thermal modelling of a novel Li-ion battery pack thermal management system. Appl Energy 181:1–13
32. Hu X, Feng F, Liu K, Zhang L, Xie J, Liu B (2019) State estimation for advanced battery management: key challenges and future trends. Renew Sustain Energy Rev 114:109334
33. Hannan MA, Lipu MH, Hussain A, Mohamed A (2017) A review of lithium-ion battery state of charge estimation and management system in electric vehicle applications: challenges and recommendations. Renew Sustain Energy Rev 78:834–854
34. Liu K, Tang X, Widanage WD (2020) Light-weighted battery state of charge estimation based on the sigma-delta technique. IFAC-PapersOnLine 53(2):12446–12451
35. Chang W-Y (2013) The state of charge estimating methods for battery: a review. Int Sch Res Notices
36. Zheng F, Xing Y, Jiang J, Sun B, Kim J, Pecht M (2016) Influence of different open circuit voltage tests on state of charge online estimation for lithium-ion batteries. Appl Energy 183:513–525
37. Zou C, Hu X, Dey S, Zhang L, Tang X (2017) Nonlinear fractional-order estimator with guaranteed robustness and stability for lithium-ion batteries. IEEE Trans Ind Electron 65(7):5951–5961
38. Guo Y, Yang Z, Liu K, Zhang Y, Feng W (2021) A compact and optimized neural network approach for battery state-of-charge estimation of energy storage system. Energy 219:119529
39. Meng J, Luo G, Gao F (2015) Lithium polymer battery state-of-charge estimation based on adaptive unscented Kalman filter and support vector machine. IEEE Trans Power Electron 31(3):2226–2238
40. Farmann A, Sauer DU (2016) A comprehensive review of on-board State-of-Available-Power prediction techniques for lithium-ion batteries in electric vehicles. J Power Sources 329:123–137
41. Wang Y, Pan R, Liu C, Chen Z, Ling Q (2018) Power capability evaluation for lithium iron phosphate batteries based on multi-parameter constraints estimation. J Power Sources 374:12–23

42. Feng T, Yang L, Zhao X, Zhang H, Qiang J (2015) Online identification of lithium-ion battery parameters based on an improved equivalent-circuit model and its implementation on battery state-of-power prediction. J Power Sources 281:192–203
43. Tang X, Liu K, Liu Q, Peng Q, Gao F (2021) Comprehensive study and improvement of experimental methods for obtaining referenced battery state-of-power. J Power Sources 512:230462
44. Pei L, Zhu C, Wang T, Lu R, Chan C (2014) Online peak power prediction based on a parameter and state estimator for lithium-ion batteries in electric vehicles. Energy 66:766–778
45. Tang X, Gao F, Liu K, Liu Q, Foley AM (in press) A balancing current ratio based state-of-health estimation solution for lithium-ion battery pack. IEEE Trans Ind Electron. https://doi.org/10.1109/TIE.2021.3108715
46. Hu X, Yuan H, Zou C, Li Z, Zhang L (2018) Co-estimation of state of charge and state of health for lithium-ion batteries based on fractional-order calculus. IEEE Trans Veh Technol 67(11):10319–10329
47. Tang X, Liu K, Lu J, Liu B, Wang X, Gao F (2020) Battery incremental capacity curve extraction by a two-dimensional Luenberger–Gaussian-moving-average filter. Appl Energy 280:115895
48. Lu J, Wu TP, Amine K (2017) State-of-the-art characterization techniques for advanced lithium-ion batteries. Nat Energy 2(3)
49. Waag W, Sauer DU (2013) Adaptive estimation of the electromotive force of the lithium-ion battery after current interruption for an accurate state-of-charge and capacity determination. Appl Energy 111:416–427
50. Stroe D-I, Schaltz E (2019) Lithium-ion battery state-of-health estimation using the incremental capacity analysis technique. IEEE Trans Ind Appl 56(1):678–685
51. Xiong R, Li L, Tian J (2018) Towards a smarter battery management system: a critical review on battery state of health monitoring methods. J Power Sources 405:18–29
52. Li Y, Liu K, Foley AM, Zülke A, Berecibar M, Nanini-Maury E, Van Mierlo J, Hoster HE (2019) Data-driven health estimation and lifetime prediction of lithium-ion batteries: a review. Renew Sustain Energy Rev 113:109254
53. Meng J, Cai L, Stroe D-I, Luo G, Sui X, Teodorescu R (2019) Lithium-ion battery state-of-health estimation in electric vehicle using optimized partial charging voltage profiles. Energy 185:1054–1062
54. Feng F, Hu X, Liu K, Che Y, Lin X, Jin G, Liu B (2020) A practical and comprehensive evaluation method for series-connected battery pack models. IEEE Trans Transp Electrification 6(2):391–416
55. Liu K, Li K, Peng Q, Guo Y, Zhang L (2018) Data-driven hybrid internal temperature estimation approach for battery thermal management. Complexity 2018
56. Hu C, Youn BD, Chung J (2012) A multiscale framework with extended Kalman filter for lithium-ion battery SOC and capacity estimation. Appl Energy 92:694–704
57. Zou C, Manzie C, Nešić D, Kallapur AG (2016) Multi-time-scale observer design for state-of-charge and state-of-health of a lithium-ion battery. J Power Sources 335:121–130
58. Dong G, Wei J, Chen Z (2016) Kalman filter for onboard state of charge estimation and peak power capability analysis of lithium-ion batteries. J Power Sources 328:615–626
59. Feng F, Teng S, Liu K, Xie J, Xie Y, Liu B, Li K (2020) Co-estimation of lithium-ion battery state of charge and state of temperature based on a hybrid electrochemical-thermal-neural-network model. J Power Sources 455:227935
60. Wei Z, Zhao J, Ji D, Tseng KJ (2017) A multi-timescale estimator for battery state of charge and capacity dual estimation based on an online identified model. Appl Energy 204:1264–1274
61. Gao Y, Liu K, Zhu C, Zhang X, Zhang D (2022) Co-estimation of state-of-charge and state-of-health for lithium-ion batteries using an enhanced electrochemical model. IEEE Trans Ind Electron 69(3):2684–2696
62. Moura SJ, Chaturvedi NA, Krstić M (2014) Adaptive partial differential equation observer for battery state-of-charge/state-of-health estimation via an electrochemical model. J Dyn Syst Meas Control 136(1)

63. Zhang X, Gao Y, Guo B, Zhu C, Zhou X, Wang L, Cao J (2020) A novel quantitative electro-chemical aging model considering side reactions for lithium-ion batteries. Electrochim Acta 343:136070
64. Tanim TR, Rahn CD, Wang C-Y (2015) State of charge estimation of a lithium ion cell based on a temperature dependent and electrolyte enhanced single particle model. Energy 80:731–739
65. Klass V, Behm M, Lindbergh G (2015) Capturing lithium-ion battery dynamics with support vector machine-based battery model. J Power Sources 298:92–101
66. Feng X, Weng C, He X, Han X, Lu L, Ren D, Ouyang M (2019) Online state-of-health estimation for Li-ion battery using partial charging segment based on support vector machine. IEEE Trans Veh Technol 68(9):8583–8592
67. Song Y, Liu D, Liao H, Peng Y (2020) A hybrid statistical data-driven method for on-line joint state estimation of lithium-ion batteries. Appl Energy 261:114408

Chapter 5
Data Science-Based Battery Operation Management II

This chapter focuses on the data science-based management for another three key parts during battery operations including the battery ageing/lifetime prognostics, battery fault diagnosis, and battery charging. For these three key parts, their fundamentals are first given, followed by the case studies of deriving various data science-based solutions to benefit their related operation management.

5.1 Battery Ageing Prognostics

Battery would inevitably degrade during the operation period, further affecting its safety and efficiency. In this context, it is of extreme importance for developing effective data science-based solutions to benefit battery ageing/lifetime prognostics. This section would first introduce battery ageing mechanism and related stress factors, then the framework of performing Li-ion battery ageing prediction with data science is given, followed by two case studies of deriving different data science-based solutions to achieve cyclic ageing and lifetime predictions of Li-ion battery.

5.1.1 Ageing Mechanism and Stress Factors

5.1.1.1 Li-Ion Battery Ageing Mechanism

Li-ion battery ageing is a complicated and long-term procedure. Understanding mechanisms of battery ageing is the prerequisite for designing data science-based tools and methodologies for battery ageing prognostics. Research has been conducted to analyse the essential reasons for battery degradation [1, 2]. The most effective and ideal way is to translate battery ageing knowledge into a mathematical form. We

© The Author(s) 2022
K. Liu et al., *Data Science-Based Full-Lifespan Management of Lithium-Ion Battery*, Green Energy and Technology,
https://doi.org/10.1007/978-3-031-01340-9_5

will give a brief introduction of the most common battery ageing mechanisms in this subsection.

The main degradation mode in Li-ion battery can be divided into three categorizations, that is, the loss of lithium inventory (LLI), the loss of active material (LAM) in the electrodes, and the increase of internal resistance. The side reactions in Li-ion battery will consume lithium inventory, and then only fewer Li-ion is available for the charging or discharging process. The related side reactions include: electrolyte decomposition reactions or lithium plating, the formulation of the solid electrolyte interface (SEI) on the surface of the graphite negative electrode. Regarding LAM, the electrodes structure changes with the volume of active materials during cycling. Then, the mechanical stress is induced by the above process, which causes particle cracking and thus reduces the density of lithium storage. In addition, the chemical decomposition and dissolution reactions of transition metals into the electrolyte and SEI modification also have an effect on the LAM. The internal resistance increase will be caused by the formation of the SEI and the loss of electrical contact inside the porous electrode [3].

Li-ion battery deteriorates in both cycling and storage conditions, which implies the cycling ageing and calendar ageing of battery [4, 5]. Generally, mechanical strains of the active materials in the electrode or lithium plating are the main reasons for cycling ageing, while the evolution of passivation layers is the dominant ageing mechanism of calendar ageing [6]. This chapter will not distinguish the cycling and calendar ageing of a battery in detail, and readers may refer to [7] for more information.

5.1.1.2   The Stress Factors for Li-Ion Battery Degradation

A variety of external factors affect battery ageing process. Besides the high SoC and temperature, overcharge/discharge, current rate, and cycling depths all have influences on battery degradation [8]. Notably, those factors are not linearly characterized with battery health status, which significantly complicates the battery ageing.

The related stress factors are outlined in Fig. 5.1 and will be introduced one by one in the following content.

(1) *High temperatures*: Extremely high temperature may easily lead to "thermal runaway" of a cell, which is an ultimate threat to battery operation management. Moreover, high temperature also accelerates the side reactions, for example, the SEI layer grows faster on the anode, and then the LLI and internal resistance are increased. Additionally, the metal dissolution from cathode and electrolyte decomposition also speeds up the LAM and LLI.

(2) *Low temperatures*: Low temperature slows down the Li-ion transport in the electrodes and electrolyte. Attempting to fast charging at low temperature may lead to the crowding of Li-ions. Thus, LLI appears with lithium plating of graphite, which eventually causes the growth of lithium dendrites and further penetrates the separator and short circuit inside a battery.

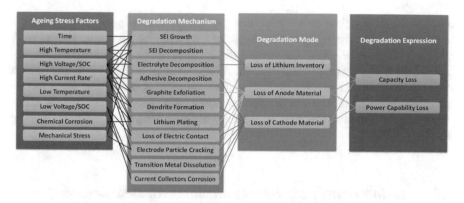

Fig. 5.1 Battery ageing stress factors and the related degradation modes, reprinted from [7], with permission from Elsevier

(3) **Overcharge/discharge**: In the case of overcharging, none active Li-ion is available on the cathode and also not enough room is left for Li-ion in the anode. Then, the structure of cathode material irreversibly changes with overdelithiated. Afterwards, the active material decomposition and the dissolution of transition metal ions occur inside the battery. Thus, when the battery undergoes an overcharging process, the electrolyte is decomposed and the total resistance is significantly increased. Considering the heat generated by side reactions at electrodes, overcharging a battery usually generates a lot of heat. Additionally, there is an abnormal increase in the anode potential, followed by the anodic dissolution of copper current collector and the formation of copper ions. Therefore, a risk of internal short circuit exists, because the reverse reactions can form copper dendrites.

(4) **High currents**: A large current will cause localized overcharge and overdischarge inside the cell, and a high current usually accompanies more heat generated. In this thread, Li-ion battery with organic electrolytes can prone the rapid temperature increase in comparison with water-based electrolytes. In addition, fast charging will accelerate the metallic Li-plating as the graphite presents a very limited capability to accept Li-ion in this condition.

(5) **Mechanical stresses**: The mechanical stresses of cell come from different aspects, such as electrode material expansion, gas evolution in mechanically constrained cells, and external loading during service. The highest stress inside cell comes from the electrode particles near the separator, which leads to the risk of cracking and fracture. Once exceeding a certain limitation, a material failure occurs in the electrode. In this condition, the performance of the cell significantly deteriorates.

For a large-scaled battery-based energy storage system, battery power flow needs to be managed according to the requirement of the load. Moreover, battery operation management system has to ensure the safety and manage the lifetime of battery. For example, the charge or discharge of battery has to be suspended, once a battery reaches its fully states; an optimal charging profile can be planned to achieve a better trade-off between the battery capacity fade and charging time. Hence, fully understanding of the impact factors is essential to establish a reliable data science-based battery ageing prognostic method. A large amount of highly quantity dataset is necessary for battery ageing prediction methods with big data. Some stress factors are more important than others in a specific application, which means that the factor has the greatest impact should be considered when designing the experimental testing matrix.

5.1.2 Li-Ion Battery Lifetime Prediction with Data Science

Li-ion battery lifetime prediction with big data will be introduced and discussed in this subsection. The recently proposed data science methods for battery lifetime prognostics are overviewed at the beginning. Afterwards, machine learning (ML) methods for battery lifetime prediction are introduced. Two case studies of our previous work using modified Gaussian process regression (GPR) and hybrid data science model will be detailed, and the conclusions are summarized at the end of this subsection.

5.1.2.1 Overview of Battery Lifetime Prognostics Methods

The battery lifetime prediction is a key part to indicate the remaining service time of battery. The remaining useful lifetime (RUL) is typically reached when the predefined battery degradation index arrives at a specific threshold. RUL can be obtained by calculating the estimated lifespan of a training unit minus the current life position. The relationship between the diagnosis and the prognostics is shown in Fig. 5.2.

In the framework of Fig. 5.2, two typical data science-based models are often used for battery lifetime prediction, which is analytical models and ML models. Analytical model needs the development of an ageing model by fitting a mathematical function to describe a large set of measurement datasets under laboratory conditions. On the contrary, ML model can directly learn from the ageing dataset itself to predict battery lifetime [9].

This subsection will mainly introduce the analytical model with data fitting. Analytical model tries to use a mathematical function representing the connections of a battery and its service time or cycling number. It can be divided into two categories: semi-empirical lifetime prediction model and empirical ageing model with filtering. Semi-empirical model is an open-loop approach, where model performance largely depends on the fitting process of the ageing dataset. Once the model is constructed, its parameters cannot be varied anymore. Meanwhile, the empirical model with filtering is able to be updated according to the new dataset with the benefit of a closed-loop structure.

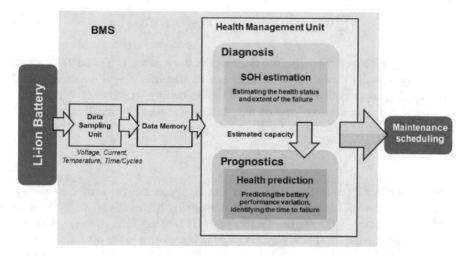

Fig. 5.2 Data science-based battery health diagnostics and prognostics framework, reprinted from [7], with permission from Elsevier

(1)   *Semi-empirical life prognostics models*

In order to establish a mathematical expression describing the battery lifetime performance, semi-empirical lifetime prediction models have been used to directly capture the relationship between ageing stress factors and battery health status. Mostly, the semi-empirical model is constructed by interpolating and fitting dataset collected from a specific experimental test [10]. For a better accuracy, those datasets should come from a wide range of operation conditions. In reality, it is rather difficult to consider the effects of all the related factors as previously described. Therefore, only the most important factors are considered for simplification.

Lots of studies build the cycling and calendar ageing model of battery independently, and the combination of these two models can generate the prediction under dynamic load profiles [11]. In order to obtain the models, the cells are cycled or stored under a well-designed test condition so that the influence of different ageing factors, such as temperature, SoC and current rates, can be deeply investigated. Afterwards, the capacity loss is calculated as the function of time, cycle numbers, and Ah-throughput. Ah-throughput is the amount of charge from one electrode to the other. The selection of fitting equations relies on the measured battery capacity trajectory. In this way, the parameters of the lifetime model are determined by fitting a large amount of ageing dataset. Once the model is constructed, the parameters are unable to be changed. Ah-throughput and current cycle number should be registered as the input for the battery future capacity prediction during operation. In addition, the battery usage conditions and loads are fed to the model if the battery lifetime is predicted.

In calendar ageing, the capacity loss is usually proportional to power law relation with time, which can be expressed as follows [12],

$$Q_{\text{loss}}^{\text{Cal}}(t) = Q(t) - Q(0) = k_{\text{Cal}}(T, \text{ SoC}) \cdot t^{Z_{\text{Cal}}} \tag{5.1}$$

where $Q_{\text{loss}}^{\text{Cal}}$ is battery capacity loss during calendar ageing, $Q(t)$ and $Q(0)$ stand for battery capacity values at time point t and its start life, Z_{Cal} is a dimensionless constant, k_{Cal} represents a stress factor related to battery temperature T and SoC. In general, the dependence of k_{Cal} on temperature T can be empirically captured by using the Arrhenius equation as [13]:

$$k_{\text{Cal}} = A \cdot e^{-E_a/RT} \tag{5.2}$$

where A denotes a pre-exponential factor and E_a represents the activation energy. Additionally, the SoC dependence on calendar ageing lifetime is typically fitted by the linear functions [14], exponential functions [15], or the Tafel equation [12].

On the other hand, the battery cycling ageing is sensitive to the operation profile. Thus, the prediction of cycle life is more complicated than calendar life prediction because more variables are involved in this condition. These factors include temperature, cycle number/time, charge/discharge current rate, cycling voltage range, and average SoC during cycling. Cycle number is often used as a measure of time for cycle lifetime modelling. One common cycling ageing model is shown in Eq. (5.3), which uses power law to express the capacity loss as:

$$Q_{\text{loss}}^{\text{Cyc}}(L) = Q(L) - Q(0) = k_{\text{Cyc}}(T, I, \text{ DoD}) \cdot L^{Z_{\text{Cyc}}} \tag{5.3}$$

where $Q_{\text{loss}}^{\text{Cyc}}$ reflects battery capacity loss during cycling ageing, which means the overall capacity difference over time/cycles, L is either cycle number or Ah-throughput, k_{Cyc} represents the effects of the ageing factors on capacity degradation, I is the cycling current. DoD is the depth of discharge during cycling. The parameters in Eq. (5.3) can be fitted from experimental dataset. Arrhenius equation can be used to empirically account for the effect of temperature. In addition, the current rate and depth-of-discharge (DoD) dependency on cycling ageing can also be expressed using exponential or polynomial functions. For example, the polynomial functions can be used to describe the capacity fade with DoD and cycling number as follows:

$$Q_{\text{loss}}^{\text{Cyc}}(L) = \sum_{i=0, j=0}^{n, m} \left(a_i \cdot L^i + b_j \cdot \text{DoD}^j \right) \tag{5.4}$$

where L is the cycle number, a_i and b_j are the fitting coefficients, n means the order of L-factor, and m is the order of DoD-factor.

(2) *Empirical ageing model with filtering*

Empirical ageing models with filtering are able to update their parameters once new data is available. A preliminary prediction model is constructed first by fitting the

Table 5.1 Models and filters used for the battery RUL prediction, reprinted from [7], with permission from Elsevier

Model equation	Filter	References
$c_k = a_1 - a_2 \cdot k$	Fixed-lag Multiple Model PF	[16]
$c_k = 1 - a_1 \big[1 - \exp(a_2 \cdot k) \big] - a_3 \cdot k$	Interacting Multiple Model PF Gauss–Hermite PF Fixed-lag Multiple Model PF	[17] [18] [16]
$c_k = a_1 \cdot \exp(a_2 \cdot k)$	PF Spherical Cubature PF	[19] [20]
$c_k = a_1 \cdot \exp(a_2 \cdot k) + a_3 \cdot k^2 + a_4$	PF	[21]
$c_k = a_1 \cdot \exp(a_2 \cdot k) + a_3 \cdot \exp(a_4 \cdot k)$	Bayesian Monte Carlo Unscented Kalman Filter Unscented PF Heuristic Kalman optimized PF Interacting multiple model PF Gauss–Hermite PF Interacting multiple model PF	[22] [23] [24] [25] [17] [26]
$c_k = a_1 \cdot k^3 + a_2 \cdot k^2 + a_3 \cdot k + a_4$	PF	[27]

experimental data to a suitable function describing the capacity degradation. The generally used linear, exponential, and polynomial functions are listed in Table 5.1. Then, the filtering methods should be applied to update the parameters of the model during battery degradation process, when the new measurement arrives. In this way, the models can be adjusted to provide a more accurate RUL prediction.

From Table 5.1, we can find that Kalman filter (KF), particle filter (PF), and their variants can enable the dynamic update of the prediction model. The observation is applied to estimate and update the parameters according to the form of a probability density function (PDF) in Bayesian inference. The filter can be chosen by the dynamic of the system and the noise distributions. For example, KF is often a good choice for linear system with Gaussian noise. A linear capacity fade model with KF is used to predict the RUL of the valve-regulated lead–acid battery in [28]. Since the Li-ion battery fading process is often nonlinear, variant KF like extended KF and unscented KF can be used to address this issue. However, for the KF family, the state space PDF is still Gaussian distributed at each iteration. In reality, the errors of RUL may not come from multiple sources for the data acquisition and transmission. Thus, the noise cannot always follow the Gaussian distribution. In this case, the KF algorithms may cause the divergence of the prediction.

For the purpose of solving non-Gaussian distribution with a nonlinear system, PF is more widely used in this thread. PF belongs to the sequential Monte Carlo method which utilizes the Bayesian inference and the importance sampling method [29]. The Bayesian update is able to deal with the particles that have the probability information of unknown parameters. When new measurement is arrived, the posterior from the previous step acts as the prior information for the current step. Therefore, the parameters are updated by multiplying it with the likelihood [30]. In the area of

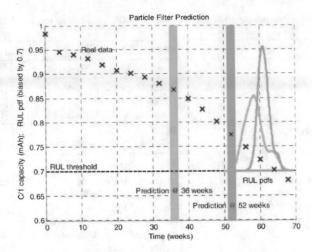

Fig. 5.3 Schematic illustration of RUL prediction with PF, reprinted from [7], with permission from Elsevier

battery RUL prediction, numerous studies related to PF and its variants have been carried out [31].

The sum of the impedance parameters shows a linear correlation with battery capacity in [32]. An impedance growth function is used to describe the battery ageing behaviour, which is combined with the PF framework to implement the battery RUL prediction in Fig. 5.3. It shows that the prediction accuracy is improved by increasing the considered dataset. It should be noted that the performance of the empirical ageing model with filtering is highly dependent on the fitted prediction model. Only one model may not enough for the complex ageing behaviour. Thus, Ref. [16] proposes two empirical models for the prediction of the Li-ion battery degradation.

5.1.2.2 Comparisons of Battery Lifetime Prediction with Data Science

The data science methods for RUL prediction have some differences in complexity, prediction accuracy, and the ability to produce confidence intervals. The main features of data science-based battery lifetime prediction methods are summarized in Table 5.2. The prediction accuracy of semi-empirical lifetime models relies on the developed mathematical function. However, this type of lifetime model has open-loop nature, which is established according to a large amount of ageing dataset from laboratory test. These models have a lower computational burden and can be easily applied to the hardware for battery operation management. It should be noted that the methods do not have a recalibration mechanism because of an open-loop structure. The prediction accuracy is also related to the amount of dataset. The empirical ageing models with filtering belong to the closed-loop type. Thus, they can adaptively adjust to the desired prediction results online. In addition, the parameters of the model can

Table 5.2 A comparison of battery lifetime prediction methods, reprinted from [7], with permission from Elsevier

		Advantages	Disadvantages
Analytical model with data fitting	Semi-empirical model	• Easy to be established; • Simple parameters extraction process; • Low computational burden; • Easy for online application	• A time-consuming and costly laboratory test procedure; • The long-term battery degradation test must be well-designed; • Comparatively poor generalizability
	Empirical ageing model with filtering	• Small amount of dataset is required; • Estimation is corrected according to the measurement	• Relatively high computing burden for real-time application
ML	Non-probabilistic	AR model	• Simple structure; • Easy parameter identification; • Easy for implementation	• Limited ability due to linear regression; • Poor generalization ability; • Not suitable for long-term prediction
		NN	• Powerful nonlinear ability; • Long-term prediction ability because of the recurrent network; • High prediction accuracy	• Lead to overfitting; • Cannot predict uncertainties; • Highly rely on the training dataset
		SVM	• High accuracy; • Low prediction time; • Robust to outliers	• High computational burden; • Poor uncertainty management ability; • Need cross-validation for the hyperparameters
	Probabilistic	GPR	• Enable the uncertainty level; • Nonparametric; • Flexible	• High related with kernel functions; • High computational cost

(continued)

Table 5.2 (continued)

		Advantages	Disadvantages
	RVM	• Generate PDF directly; • Nonparametric; • High sparsity; • Avoid cross-validation procedure	• Plenty data is need for modelling; • Large time and memory are required for the training; • Easy to trap into local optimization; • Overfitting

also be updated during the operation for tuning the prediction. However, the structure of those models may limit the capability of the prediction under more complex battery ageing conditions. In this way, we recommend to employ hybrid approaches. For example, a lifetime model can be combined with an adaptive filter to update the parameters of the model for a more reliable prediction. In this case, the model can be used for real battery degradation prediction in different cycling conditions.

ML methods do not rely on any explicit mathematical model, and mostly the information from the historical test dataset is utilized to describe battery ageing behaviour. The non-probabilistic methods cannot give the uncertainty level for prediction at each estimated point. However, considering the uncertainty from measurements, state estimation algorithm, and future cycling profile, the uncertainty level is critical for battery lifetime prediction. Thus, the probabilistic methods with the ability to produce PDF are able to predict the results and also give the confidence bounds. We recommend probabilistic ML methods in the battery lifetime estimation. Another issue here is that most existing researches validate their methods with the same dataset in the model training phase. There exists a question about the generalization of those models in real applications, in which the cycling profile is completely different. Of course, the generalization of the models can be improved by training under more complex ageing conditions. Moreover, suitable structures and parameters should be investigated to develop a self-adaptive battery lifetime predictor in future [33].

In order to accelerate the development and optimization of battery technologies, some methods are proposed to accurately predict the lifetime of battery in a very early stage. Reference [34] tries to solve this issue using lasso and elastic-net regression based on a comprehensive training dataset that includes 124 commercial LFP/graphite cells. The best regression model can predict the lifetime for 90.9% of the tested cell within the first 100 cycles. Moreover, the classification model could classify cells with the first five cycles with the error of 4.9%. This work proves the great potential of applying ML techniques for battery lifespan prediction. Through coupling migration concept into GPR, a migrated GPR-based data science solution is designed in [35] to predict battery future two-stage ageing trajectory, while the knee point effect can be considered just using a small portion of starting ageing data.

Table 5.3 Cyclic ageing test matrix of tested cells, reprinted from [36], with permission from IEEE

	Cyclic DoD [%]	Temperature [°C]	Charge current rate	Discharge current rate
Case 1	100	35	C/3	1C
Case 2	50	45	C/3	1C
Case 3	50	35	C/3	1C
Case 4	100	45	C/3	1C
Case 5	80	35	C/3	1C
Case 6	80	45	C/3	1C

5.1.3 Case 1: Li-Ion Battery Cyclic Ageing Predictions with Modified GPR

In this case study, a data science-based solution by devising the modified GPR is developed to predict the future capacity of Li-ion battery with the consideration of various cyclic cases (temperatures and DoDs) [36]. Specifically, a GPR-based data science model structure is first proposed by involving inputs of cyclic battery temperature and DoD. Then after coupling the typical Arrhenius law and empirical polynomial equation with the compositional kernel, a novel GPR model is derived to integrate both electrochemical and empirical elements of battery ageing. This is the first known data science application by constructing GPR's kernel function with electrochemical and empirical knowledge of battery ageing for future cyclic capacity predictions.

5.1.3.1 Cyclic Ageing Dataset

Table 5.3 details the cyclic ageing test matrix of tested cells with the same middle-SoC of 50% but under various cycling DoDs (50%, 80%, and 100%) and temperatures (35 and 45 °C). More detailed experimental information can be found in [36], which is not repeated here due to space limitations. According to this test matrix, battery capacity dataset under six cyclic cases can be obtained, as shown in Fig. 5.4. To ensure the derived model could study enough mapping mechanism, four cases including Case 1, Case 2, Case 3, and Case 4 are utilized as the training dataset, while other two cases (Case 5 and Case 6) are adopted for the validation purpose.

5.1.3.2 Model Structure for Battery Cyclic Capacity Prediction

According to this dataset, a GPR model structure through involving a series of capacity terms $C_{bat}(t - i)$ (here i represents the previous time period), cyclic temperature T_{cyclic} and DoD DOD_{cyclic} is designed to perform future cyclic capacity prediction, as illustrated in Fig. 5.5.

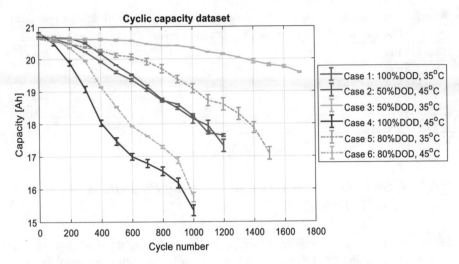

Fig. 5.4 Cyclic capacity dataset under various DoD and temperature conditions, reprinted from [36], with permission from IEEE

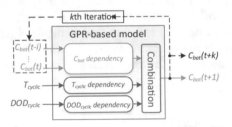

Fig. 5.5 Model structure for battery cyclic capacity prediction, reprinted from [36], with permission from IEEE

In the training process, after the combination of training dataset (here four cases are included), model's input vector and output vector are $\left[C_{\text{bat}}(t-i), \ldots, C_{\text{bat}}(t), T_{\text{cyclic}}, \text{DOD}_{\text{cyclic}}\right]$ and $C_{\text{bat}}(t+1)$, respectively. Both T_{cyclic} and $\text{DOD}_{\text{cyclic}}$ are constant for each specific cycling case. After training GPR model, both one-step $C_{\text{bat}}(t+1)$ and multi-step $C_{\text{bat}}(t+k)$ predictions are performed for Case 5 and Case 6. To capture battery future multi-step capacity, a recursive process by using the previously predicted capacity value as next input point for further predicting a new capacity point under same T_{cyclic} and $\text{DOD}_{\text{cyclic}}$ is adopted. As the previous and current capacity points are also involved, this model structure has an ability to consider battery ageing trend of different cycling cases.

5.1.3.3 Modified GPR

As T_{cyclic} and $\text{DOD}_{\text{cyclic}}$ are two key elements for determining battery cyclic ageing dynamics, their influence is thus needed to be considered carefully. To this end, an attempt has been done here to modify GPR's kernel for developing a novel data science model (labelled as Model B) that could take the electrochemical or empirical elements of Li-ion battery ageing into account. To be specific, the related components within GPR's kernel are modified separately to reflect cyclic temperature, DoD, and battery capacity.

Temperature dependency: For battery cyclic degradation, the Arrhenius equation $f_{\text{Arr}}(T)$, which shows that battery side reaction will decrease with reduced temperature exponentially, has been reported in numerous literatures [37] to reflect temperature effect as:

$$f_{Arr}(T) = a \cdot \exp(-E_A/RT) \tag{5.5}$$

where a is a pre-exponential parameter. R means the ideal gas constant. E_A represents the activation energy of electrochemical reactions. T stands for battery operational temperature.

According to the Arrhenius equation, a component $k_{T\text{cyc}}(x_T, x'_T)$ related to T_{cyclic} within GPR's kernel could be modified with the similar exponential form as:

$$k_{\text{Tcyc}}(x_T, x'_T) = l_T \cdot \exp\left(-\frac{1}{\sigma_T}\left\|\frac{1}{x_T} - \frac{1}{x'_T}\right\|\right) \tag{5.6}$$

where l_T and σ_T are two hyperparameters. It should be known that this temperature dependency is described by an isotropic form to reflect the relevance degree between outputs generated by the difference of temperatures x_T and x'_T. In this context, GPR model successfully couples the Arrhenius law to capture temperature dependency.

DoD dependency: Based upon numerous experimental conclusions [38], the effects of DoD on battery cycling degradation usually show a polynomial or linear trend. That is, DoD dependency can be empirically captured by using the polynomial equation. In this context, a specific component $k_{\text{DOD}}(x_{\text{DOD}}, x'_{\text{DOD}})$ is modified with the polynomial form to describe DoD dependency as:

$$k_{\text{DOD}}(x_{\text{DOD}}, x'_{\text{DOD}}) = \left(l_D \cdot x_{\text{DOD}}^T x'_{\text{DOD}} + c_D\right)^{d_D} \tag{5.7}$$

where l_D, c_D, and d_D are three related hyperparameters. It should be noted that this component is a none stationary kernel, which could benefit computation effort as the none stationary kernel generally needs a small number of data to train.

Capacity dependency: To describe dependency of battery capacity, a squared exponential (SE) kernel with hyperparameters l_c, σ_c is used to describe the difference of capacity terms x_c and x'_c as:

$$k_{C\text{bat}}(x_c, x'_c) = l_c^2 \exp\left(-\sum_{c=1}^{i+1} \frac{\|x_c - x'_c\|^2}{2\sigma_c^2}\right) \quad (5.8)$$

At this point, all components of GPR's kernel have been formulated to consider the electrochemical or empirical knowledge of Li-ion battery ageing. Then a novel modified kernel for "Model B" is formulated as:

$$
\begin{aligned}
k_{\text{modified}}(x, x') &= k_{C\text{bat}}(x_c, x'_c) \cdot k_{T\text{cyc}}(x_T, x'_T) \cdot k_{\text{DOD}}(x_{\text{DOD}}, x'_{\text{DOD}}) \\
&= l_f^2 \cdot \exp\left(-\frac{1}{\sigma_T}\left\|\frac{1}{x_T} - \frac{1}{x'_T}\right\|\right) \cdot \left(x_{\text{DOD}}^T x'_{\text{DOD}} + c_D\right)^{d_D} \cdot \\
&\quad \exp\left(-\sum_{c=1}^{i+1} \frac{\|x_c - x'_c\|^2}{\sigma_c^2}\right)
\end{aligned}
\quad (5.9)
$$

where $x = (x_c, x_T, x_{\text{DOD}})$. Based upon the model structure in Fig. 5.5, x_c is $[C_{\text{bat}}(t-i), \ldots, C_{\text{bat}}(t)]$, x_T is T_{cyclic}, x_{DOD} is DOD_{cyclic}. l_f, σ_T, c_D, d_D and σ_C are five related hyperparameters.

5.1.3.4   Results and Discussions

Next, the performance of modified GPR is explored based on real cyclic dataset. Figure 5.6 illustrates its training results for four cases. Through deriving the modified kernel to consider battery electrochemical and empirical degradation knowledge, modified GPR could well capture battery capacity ageing dynamics. Quantitatively, the maximum ME, MAE, and RMSE of all these cases are just 0.1689, 0.0557, and 0.0790 Ah, indicating the satisfactory fitting ability can be achieved by using modified GPR.

After the training process, both one-step and multi-step tests are carried out to explore the extrapolative prediction performance of modified GPR. According to the one-step prediction results as shown in Fig. 5.7, the well-trained model could capture battery cyclic capacity ageing trends for both Case 5 and Case 6, as indicated by the satisfactory matches among output points and real capacity data. Table 5.4 illustrates its corresponding performance indicators. Not surprisingly, the ME, MAE, and RMSE for one-step prediction of these two cases are all within 0.15Ah, indicating that modified GPR could provide highly accurate performance for one-step cyclic capacity prognostics.

Next, the evaluation of multi-step prediction of modified GPR is carried out. As illustrated in Fig. 5.8, the predicted capacities also well match the real data, which

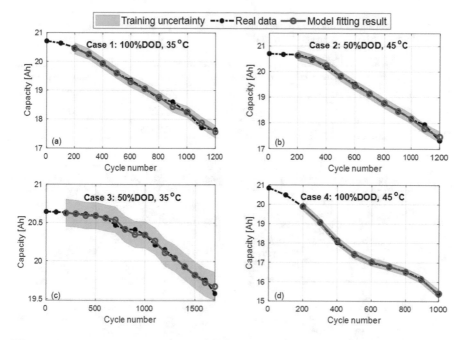

Fig. 5.6 Training results by using modified GPR for each cyclic cases from training dataset, reprinted from [36], with permission from IEEE

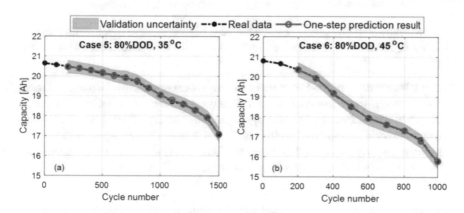

Fig. 5.7 One-step prediction results by using modified GPR for testing dataset, reprinted from [36], with permission from IEEE

implies that satisfactory multi-step prediction accuracy can be obtained by using the designed model. For Case 5 in Fig. 5.8a, although a few mismatches happen at the large local fluctuation conditions, the global capacity trend could be also well captured. Here the MAE and RMSE are just 0.0680 and 0.0873 Ah. Similarly, more

Table 5.4 Performance indicators for one-step and multi-step prediction results by modified GPR, reprinted from [36], with permission from IEEE

Testing cases	Case 5		Case 6	
Prediction types	One-step	Multi-step	One-step	Multi-step
ME [Ah]	0.1475	0.2576	0.0751	0.2004
MAE [Ah]	0.0447	0.0680	0.0273	0.0512
RMSE [Ah]	0.0598	0.0873	0.0355	0.0771

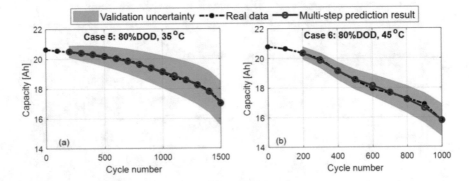

Fig. 5.8 Multi-step prediction results by using modified GPR for testing dataset, reprinted from [36], with permission from IEEE

efficient multi-step prediction results can be obtained in Case 6. Quantitatively, the ME, MAE, and RMSE values are all within 0.2004 Ah for cyclic capacity predictions of Case 6.

Comparison with other GPR models: To further explore the performance of modified GPR model (here named as Model B), a typical SE-based GPR model (SEGM) with two hyperparameters (σ_f and σ_l) in the form of Eq. (5.10) and an automatic relevance determination (ARD)-SE-based GPR model (Model A) with four hyperparameters (σ_f, σ_T, σ_{DOD}, σ_c) in the form of Eq. (5.11) are also adopted here. Table 5.5 shows the optimized hyperparameters of these GPR models.

Table 5.5 Hyperparameters of GPR-based models, reprinted from [36], with permission from IEEE

Model types	Hyperparameters
SEGM	$\sigma_f = 0.894$, $\sigma_l = 2.036$
Model A	$\sigma_f = 0.826$, $\sigma_T = 1.813$, $\sigma_{DOD} = 2.391$,
	$\sigma_1 = 2.489$, $\sigma_2 = 1.430$
Model B	$l_f = 0.516$, $\sigma_T = 3.964$, $c_D = 4.520$, $d_D = 1.323$,
	$\sigma_1 = 5.282$, $\sigma_2 = 3.351$

$$k_{SE}(x, x') = \sigma_f^2 \exp\left(-\frac{\sum_{c=1}^{i+1}\|x_c - x_c'\|^2 + \|x_T - x_T'\|^2 + \|x_{DOD} - x_{DOD}'\|^2}{2\sigma_l^2}\right)$$

(5.10)

$$k_{ARD\,SE}(x, x') = \sigma_f^2 \exp\left[-\frac{1}{2}\left(\frac{\|x_T - x_T'\|^2}{\sigma_T^2} + \frac{\|x_{DOD} - x_{DOD}'\|^2}{\sigma_{DOD}^2}\right.\right.$$
$$\left.\left. + \sum_{c=1}^{i+1}\frac{\|x_c - x_c'\|^2}{\sigma_c^2}\right)\right]$$

(5.11)

(1) *Comparison with the Training Results*: Fig. 5.9 shows the performance indicators of training results for SEGM, Model A and Model B. Obviously, through using the GPR technique with ARD kernel and modified kernel, the training results of Model A and Model B are both better than those from SEGM. Quantitatively, after coupling GPR models with improved kernels, the related training performance can be enhanced nearly twice in comparison with the SEGM.

(2) *Comparison with the Prediction Results*: To further evaluate the multi-step prediction performance of each model type, the corresponding performance indicators for total testing dataset are compared and illustrated in Fig. 5.10. It can be seen that the ME and RMSE for Model A and Model B are within 0.71 and 0.31 Ah, which are 32.6 and 13.6% less than those of SEGM. In addition, in comparison with Model A owns the ARD-SE kernel, Model B also provides the significant improvement for multi-step cyclic capacity predictions. The RMSE here becomes 0.0835 Ah (72.3% decrease), indicating the superiority of coupling electrochemical or empirical knowledge into GPR.

Based upon the above results, Model B outperforms Model A and SEGM for both the one-step and multi-step predictions. This suggests that the data science model considering battery electrochemical or empirical knowledge is promising for predicting battery future capacities under different cycling cases.

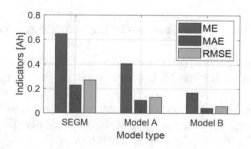

Fig. 5.9 Indicators of using different model types for all training dataset, reprinted from [36], with permission from IEEE

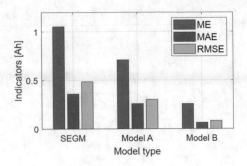

Fig. 5.10 Indicators of using different model types for multi-step prediction, reprinted from [36], with permission from IEEE

5.1.4 Case 2: Li-Ion Battery Lifetime Prediction with LSTM and GPR

In this subsection, we will introduce a long short-term memory (LSTM) and Gaussian process regression (GPR)-based hybrid data science model to predict battery future capacities and RUL during its cyclic conditions [39]. To achieve reliable future capacities and RUL prediction, three key points need to be concerned: First, the raw capacity data shows the highly nonlinear trend with regeneration phenomenon, which would significantly affect the accuracy of battery health prognosis. Second, capturing the interactions of battery capacity time-series is crucially important to understand its long-term dependencies. Third, prediction uncertainty would occur frequently and should not be ignored.

To handle these challenges, the utilized hybrid data science-based model mainly contains three parts: an empirical mode decomposition (EMD) part to decompose the raw capacity data, a LSTM submodel part to learn the long-term fading dependence of capacity, and a GPR submodel part to quantify the uncertainties of prediction results.

5.1.4.1   Hybrid Data Science Framework for Future Ageing Prediction

The framework of using this hybrid data science-based model as well as the workflow to predict future capacities and RUL of battery is shown in Figs. 5.11 and 5.12, respectively.

The utilized hybrid data science-based model can be divided into two parts. For the battery future capacities prediction, with the current and historical capacity vector $\{C_{\mathrm{bat}}(t - i), \ldots, C_{\mathrm{bat}}(t - 1), C_{\mathrm{bat}}(t)\}$ as the inputs of model, the future capacity $C_{\mathrm{bat}}(t + k)$ could be predicted through using GPR submodel and LSTM submodel to study the potential mappings of intrinsic mode functions (IMFs) and residual after using the EMD to decouple the raw capacity data. Here k and i represent the future step as well as previous step, respectively. The details of EMD technique can be

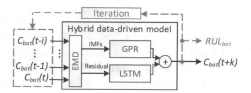

Fig. 5.11 Framework of using hybrid data science-based model to predict future capacities and RUL of battery, reprinted from [39], open access

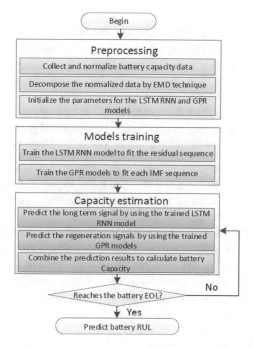

Fig. 5.12 Workflow of using the hybrid LSTM + GPR model to predict future capacities and RUL of battery, reprinted from [39], open access

found in [40] for the readers of interest. For the battery RUL prognostics, a recursive prediction process which adopts the previously predicted capacity as the next model input to further predict new capacity value is carried out iteratively until the end-of-life (EoL) of battery is reached. Then the corresponding RUL (RUL_{bat}) could be calculated. It should be known that the capacity prediction is carried out just based on the historical capacity information. Detailed prediction workflow is described as follows:

Step 1: Preprocessing: for the data preprocessing before any training processes, an efficient and simple normalization approach [41] is adopted to convert the raw capacity data C_{bat} into a normalized level C'_{bat} through using an equation:

$v' = v/C_{new}$. Here C_{new} stands for the fresh capacity value of a battery. v' and v are the data samples in C'_{bat} and C_{bat}, respectively. Then the data will be decomposed into several IMFs and a residual through adopting the EMD technique. For the hybrid model part, select the suitable kernel function (here is rational quadratic kernel [42]) for GPR. Set the structure as well as initialize the parameters for LSTM and GPR submodels.

Step 2: Models training: for a decomposed residual sequence, train the LSTM submodel to fit the residual sequence. For the obtained IMFs, train the GPR submodels to fit each IMF sequence.

Step 3: Estimation of battery future capacities: for the long-term signal part, using the well-trained LSTM submodel to predict the future residual value. For the regeneration signal part, applying the well-trained GPR models to capture the mean and covariance values of each IMF. Afterwards, the predicted battery future capacities along with the corresponding uncertainty quantification can be obtained through combining these results.

Step 4: Predicting battery RUL: calculating the battery capacity when it reaches the EoL as C_{EOL}. Repeating the prediction step until the predicted future capacity fades below C_{EOL}. Then output the predicted RUL of battery.

Following this workflow, the capacities in future cycles of battery could be estimated. Then the battery RUL can be also predicted to provide valuable information for the maintenance decision of aged battery, while the uncertainties of predicted results could be quantified accordingly.

Based upon the above-mentioned workflow, to evaluate the extrapolation performance of this hybrid data-driven model, multi-step ahead prediction tests by using different horizons of 6, 12, and 24 steps are carried out. For these tests, inputs are obtained using 10 historical capacity data up to current cycle, and the prediction is carried out at the cyclic k-step ahead of the current cycle.

5.1.4.2 Results and Discussions

Figure 5.13 shows the k-step ahead prediction results for some open-source batteries [43]. It can be seen that some short-period mismatches occur in the multi-step ahead prediction cases, which should be mainly caused by the lack of priori information for the future large local fluctuations of battery capacity. However, the predicted capacity values would gradually rematch the true test ones again due to the efficient information decomposition and the strong long-term capture abilities of the utilized hybrid model. Interestingly, as the prediction step increases, the 95% confidence range would distribute in a wider region, implying that the prediction uncertainties become larger. This is hardly surprising given that the relative long-step predictions contain much more uncertainties. Even so, the max uncertainty value is still less than $\pm 10\%$ capacity range, indicating that the prognostic results are reliable. Here the uncertainty boundaries are mainly related to the so-called scope compliance

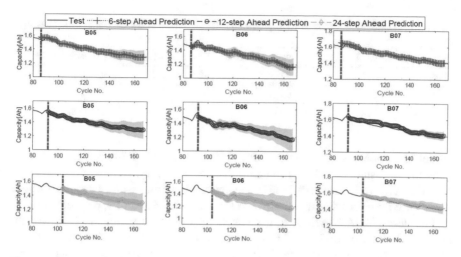

Fig. 5.13 Results of k-step ahead battery capacity predictions, reprinted from [39], open access

uncertainty, which is used to quantify "how confident" the GPR model felt when performing predictions.

Table 5.6 illustrates the performance metrics for the battery multi-step ahead prediction. According to Table 5.6, the maximum RMSE for B05, B06, and B07 are 0.0041, 0.0059, 0.0052 for the 24-step ahead prediction cases, which are 13.9, 20.4, 4% more than those of the 12-steps case, and 7.9, 15.7, and 40.5% larger than those of the 6-steps case. However, all these values are less than 0.006, indicating that satisfactory overall capacity predictions can be obtained for such cases. In the light of this, the proposed LSTM + GPR hybrid model presents a good extrapolation performance for battery multi-step ahead prediction.

According to the requirements of battery health diagnosis system, predicting the future battery RUL as early as possible with a reliable accuracy level is more meaningful for battery real-world application. In such a case, it is critically important to predicting the RUL of battery at an early cycle stage. To further investigate the recursive prediction performance and the robustness of our proposed LSTM + GPR

Table 5.6 Performance metrics for the battery multi-step ahead prediction, reprinted from [39], open access

Battery No	B05	B06	B07
RMSE (6-steps)	0.0038	0.0051	0.0037
Max error (6-steps)	0.027	0.035	0.024
RMSE (12-steps)	0.0036	0.0049	0.0050
Max error (12-steps)	0.023	0.034	0.028
RMSE (24-steps)	0.0041	0.0059	0.0052
Max error (24-steps)	0.077	0.113	0.041

Table 5.7 Quantitative results of RUL predictions for all battery cases, reprinted from [39], open access

Battery No	Actual EoL	Actual RUL	Predicted RUL	RUL uncertainties
B05	126	92	95	[90, 102]
B06	127	93	94	[90, 100]
B07	142	108	111	[105, 118]

model, the open-source batteries from NASA are tested. Table 5.7 illustrates the quantitative results for all RUL prediction cases of battery. Here, the left and right bounds of uncertainties are defined by the first and last time instant when the obtained confidence range reaches the predefined EoL value of the battery, respectively.

For these NASA batteries, to investigate the effects of various EoL values, the predefined EoLs of B05, B06, and B07 are set as 75, 66, and 77%, respectively. Figure 5.14 shows their predicted RUL results. It can be seen that for various batteries with different defined EoL values, the predicted capacities present similar trends with the real capacity curves. According to Table 5.7, the actual EoL values of B05, B06, and B07 are 126, 127, and 142 cycles, respectively. When implementing the RUL prediction at the 34th cycle (here is the first 1/5 proportion), the predicted RUL for B06 is 94, which is only 1 cycle (1.1%) later than the actual RUL. The predicted RULs for B05 and B07 are both 3 cycles (3.3 and 2.7%) later than their actual RULs. Meanwhile, all the RUL uncertainty bounds of these predictions cover the real RUL values effectively.

5.2 Battery Fault Diagnosis

To meet the endurance requirements of electric vehicles (EVs), the battery manufacturers pursue large capacity and high energy density by filling more active materials in battery manufacturing, which will make Li-ion battery more prone to faults and safety accidents [44]. In recent years, fire accidents in EVs caused by battery faults such as thermal runaways have occurred frequently. Therefore, the battery fault diagnosis has attracted considerable attention worldwide. Generally, one obvious fault named thermal runaway is caused by mechanical, electrical, or thermal abuses. These abuses will further lead to internal short circuits (ISCs) to a certain extent [45, 46]. That is, the ISC fault is the common cause of battery thermal runaway. Therefore, the early ISC fault diagnosis is important for improving battery safety.

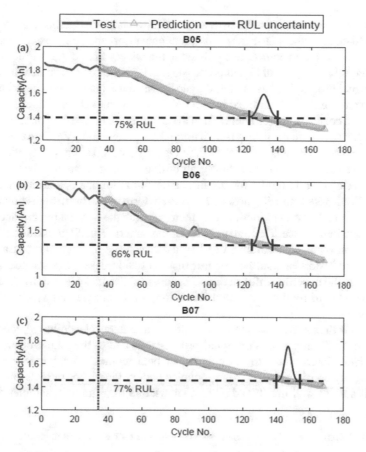

Fig. 5.14 RUL prediction results for NASA batteries. **a** B05, **b** B06, **c** B07, reprinted from [39], open access

5.2.1 Overview of Data Science-Based Battery Fault Diagnosis Methods

For battery fault diagnosis, related data science-based approaches would directly analyse and explore the battery operation data to detect faults without the requirements of accurate analytical models and expert experiences [47]. Here the battery fault detection process will be simplified without taking complex fault mechanisms and system structures into account, especially for thermal runaways being affected by different factors. However, data science-based battery fault diagnosis methods generally require a suitable preprocess stage to handle raw battery data. As fault mechanisms are usually ignored, it becomes difficult to perform faults explanations based on this method. Besides, several data science-based battery fault diagnosis methods

also present inherent drawbacks such as the requirement of numerous battery historical fault data to further cause large computational effort and complexity. In general,
the data science-based approaches utilized in the battery fault diagnosis domain can
be divided into the types of signal processing, ML, and information fusion.

For signal processing-based fault diagnosis, various signal processing techniques
are adopted to extract feature parameters of faults, such as the deviation, variance,
entropy, and correlation coefficient. Then these faults will be detected through parameter comparisons with the values from a normal state. NNs and SVMs are two popular
utilized data science tools. NN-based fault diagnosis would study the implicit logics
from a given input–output pair during an offline training stage and then generates
a nonlinear black-box model for the utilizations in an online operation phase. The
logic of SVM-based fault diagnosis is to first transform inputs into a high-dimensional
space based on the kernel functions and then to search the optimal hyperplane of this
high-dimensional space. Through treating Li-ion battery fault diagnosis as the sample
classification issue, an accurate SVM-based data science model could be trained by
using related historical data. For the information fusion-based data science method,
it would use the uncertain information of battery faults to make decisions. According
to the analyses of multiple source information, more reliable battery faults can be
detected.

Table 5.8 illustrates a comparison of these three data science-based approaches
for battery fault diagnosis. For signal process one, due to the nature of neglecting
battery dynamics, it is easy to be implemented but becomes difficult to locate battery
faults directly considering the strong-coupled battery faults. For the ML one such as
NN, it is able to well match and extract knowledge from training samples through

Table 5.8 A comparison of data science-based approaches for battery fault diagnosis

Data science-approaches	Key technology	Merits	Limitations
Signal process	Advanced signal process methods	• Easy to implement • Suitable for linear and nonlinear cases	• Difficult to detect minor and locate faults • Unsuitable for highly coupled system
NN	• NN structure • Dynamic weight adjustment	• Self-learning • Strong adaptability • Parallel process	• Easy to cause overfitting • Need massive data and long training time
SVM	Kernel functions selection	• Good generalization • Suitable for small sample cases	Low efficiency for large-scale training datasets
Information fusion	Advanced information fusion methods	Highly accurate diagnostic results	Difficult to select effective fusion methods

setting suitable parameters. However, lacking enough Li-ion battery fault data would also lead to overfitting issues. In comparison with NN, SVM would present better generalization results under the small sample cases. This could become suitable for Li-ion battery cases with limited fault data. However, the hyperparameters of kernel functions within SVMs must be well-optimized or selected for the specific battery fault diagnosis issue. To make full use of multiple sources from Li-ion batteries for improving their fault diagnosis accuracy, a reliable information fusion solution becomes essential.

5.2.2 Case: ISC Fault Detection Based on SoC Correlation

In this subsection, a data science-based battery ISC detection and diagnosis method through using the battery charging data is introduced [48]. In real applications, as battery cells are strictly screened to ensure consistency before being grouped, the characteristic parameters (such as voltage, internal resistance, SoC) between cells in series should show similar trends during charging and discharging, and these parameters have a high degree of correlation between cells. When the ISC occurs in a cell, its characteristics are quite different from those of other cells due to the extra power consumption of the ISC resistor. Therefore, the correlation between the cell with ISC and the normal cells would decrease, making the ISC can be detected from the correlation of cell parameters. When a cell is charged or discharged, its voltage is an easily detected dominant signal. However, battery voltage fluctuates under dynamic conditions, and its internal resistance is difficult to be calculated online. In this context, these two parameters become unsuitable for online ISC detection.

From several existing research [49, 50], SoC estimated by EKF presents less fluctuation even under dynamic conditions. The reason for this phenomenon is that the voltage and current of the battery are the signals that change rapidly under dynamic conditions, while the SoC changes slowly at a long-term scale. Therefore, the correlation of SoC is a competitive candidate for estimating ISC. Figure 5.15 illustrates the proposed online ISC detection based on SoC correlation. Taking three batteries in series as an example, suppose there is an ISC in Cell 3. The cells in series show almost the same SoC during charging and discharging, but the ISC cell's SoC becomes slightly different from that of normal cells. Specifically, the ISC cell exhibits a faster SoC drop due to additional power loss when discharging, and the SoC of the ISC cell would increase more slowly due to power loss when charging. In addition, the SoC difference between the normal cells and the ISC cells increases with time. Therefore, the difference in SoC can be used to detect ISC online with high sensitivity.

To reduce the influence of SoC estimation cumulative error on the correlation, a method of calculating the SoC correlation coefficient in a moving window is adopted to improve the robustness. At a given moment, the SoC correlation coefficient is calculated in a fixed period, and then it is updated in real time with the moving window. As shown in Fig. 5.15, L is the size of moving window, and r_{thr} is the set threshold. Obviously, the SoC correlation coefficient between normal cells should

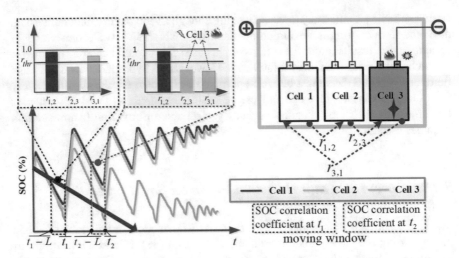

Fig. 5.15 Schematic diagram of the proposed data science-based ISC detection method, reprinted from [48], with permission from Elsevier

get close to 1, whereas that between a cell with ISC and normal cells should tend to 0. To improve this misjudgment, only the SoC correlation coefficients of a cell and two adjacent cells are less than the threshold before the ISC warning is performed.

5.2.2.1   ISC Detection Algorithm

Figure 5.16 describes the proposed ISC detection algorithm based on SoC correlation, which consists of four steps: In step 1, the voltage, total current, and temperature of each cell are collected in real time. In step 2, the first-order RC (1RC) model and EKF are used to estimate the SoC. In step 3, the SoC correlation coefficient between each two adjacent cells is calculated. Assume that the first and last cells in the series-connected batteries are adjacent. Therefore, each cell has two correlation coefficients relative to its two neighbouring cells. In step 4, the correlation coefficient is compared with a predefined threshold to determine ISC and identify the ISC cell.

Figure 5.17 illustrates the SoC estimation method based on the EKF algorithm, where $-$ represents the prior value, $+$ represents the posterior value, and the subscript k represents the time step. Specifically, the method can be described as follows:

Step 1. A priori estimation. First, a priori SoC value at time k is calculated by the ampere-hour counting method as follows.

$$\mathrm{SOC}_k^- = \mathrm{SOC}_{k-1}^+ + \frac{\Delta t_{k-1}}{C_Q \cdot 3600} I_{k-1} \tag{5.12}$$

where C_Q is the battery capacity.

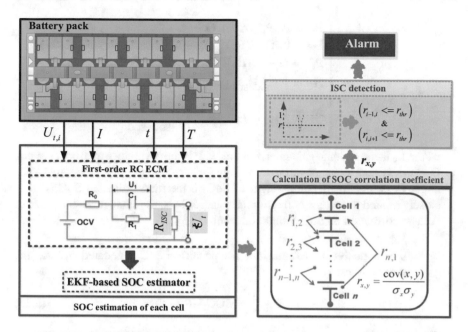

Fig. 5.16 ISC online detection algorithm flow based on SOC correlation, reprinted from [48], with permission from Elsevier

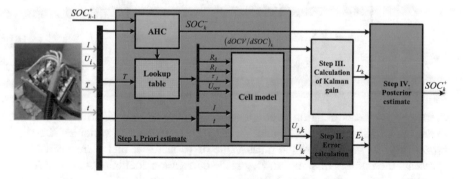

Fig. 5.17 SoC estimation process based on EKF, reprinted from [48], with permission from Elsevier

Step 2. Error calculation. The voltage error at time k (E_k) is obtained by comparing the model terminal voltage with the measured terminal voltage as follows:

$$E_k = \tilde{U}_k - U_{t,k} \qquad (5.13)$$

Step 3. Calculation of Kalman gain matrix. The Kalman gain matrix L_k at time step k is calculated as follows:

$$\begin{cases} \mathbf{P}_k^- = \mathbf{A}_k \mathbf{P}_{k-1}^+ \mathbf{A}_k^T + \mathbf{Q} \\ \mathbf{L}_k = \mathbf{P}_k^- \mathbf{H}_k^T / (\mathbf{H}_k \mathbf{P}_k^- \mathbf{H}_k^T + \mathbf{R}) \\ \mathbf{P}_k^+ = (1 - \mathbf{L}_k \mathbf{H}_k) \mathbf{P}_k^- \\ \mathbf{H}_k = \left(\frac{\partial U_{\text{ocv}}}{\partial \text{SOC}} \Big|_{\text{SOC}=\text{SOC}_k^-} \quad -1 \right) \\ \mathbf{A}_k = \left(\begin{matrix} 1 & 0 \\ 0 & \exp(-\Delta t / \tau_{1,k}) \end{matrix} \right) \Big|_{\text{SOC}=\text{SOC}_k^-} \end{cases} \tag{5.14}$$

where $\mathbf{L}_k$ is the Kalman gain matrix at time k, $\mathbf{P}$ is the covariance matrix of system, $\mathbf{Q}$ is the system noise covariance, $\mathbf{R}$ is the measurement noise covariance. The larger $\mathbf{Q}$ is, the smaller the weight of SoC in the final estimated SoC is. $\mathbf{R}$ is closely related to voltage correction. The larger $\mathbf{R}$ is, the smaller the influence of voltage correction on SoC estimation results.

Step 4. A posteriori estimation. The posteriori SoC is updated with $\mathbf{L}_k$ and error as follows:

$$\text{SOC}_k^+ = \text{SOC}_k^- + \mathbf{L}_k E_k \tag{5.15}$$

The correlation coefficient is often used to study the linear consistency between two variables. It can be expressed as follows:

$$\begin{cases} r_{X,Y} = \frac{\text{cov}(X,Y)}{\sqrt{\text{Var}(X) \cdot \text{Var}(Y)}} = \frac{\sum_{i=1}^n (X_i - \mu_X)(Y_i - \mu_Y)}{\sqrt{\sum_{i=1}^n (X_i - \mu_X)^2} \sqrt{\sum_{i=1}^n (Y_i - \mu_Y)^2}} \\ \mu_X = \frac{1}{n} \sum_{i=1}^n X_i \\ \mu_Y = \frac{1}{n} \sum_{i=1}^n Y_i \end{cases} \tag{5.16}$$

where $r_{X,Y}$ is the correlation coefficient of variables X and Y, $\text{cov}(X, Y)$ is the covariance of X and Y, $Var(X)$ and $\text{Var}(Y)$ are the variance of variables X and Y, respectively. μ_X and μ_Y are the mean value of variables X and Y, respectively. n represents the number of samples, $r_{X,Y}$ is the correlation coefficient between X and Y, and its value range is -1 to 1. When $r_{X,Y} = 0$, it means that the two variables are completely unrelated. When $r_{X,Y} < 0$, it means that the two variables are negative correlation, and when $r_{X,Y} > 0$, it means that the two variables are positive correlation.

For the convenience of calculation, Eq. (5.16) can be rewritten as follows:

$$r_{X,Y} = \frac{n \sum_{i=1}^n X_i Y_i - \left(\sum_{i=1}^n X_i \right) \left(\sum_{i=1}^n Y_i \right)}{\sqrt{n \sum_{i=1}^n X_i^2 - \left(\sum_{i=1}^n X_i \right)^2} \sqrt{n \sum_{i=1}^n Y_i^2 - \left(\sum_{i=1}^n Y_i \right)^2}} \tag{5.17}$$

To facilitate the algorithm implementation, Eq. (5.17) can be discretized as follows:

$$\begin{cases} a_k = a_{k-1} + X_k Y_k \\ b_k = b_{k-1} + X_k \\ d_k = d_{k-1} + Y_k \\ f_k = f_{k-1} + X_k^2 \\ g_k = g_{k-1} + Y_k^2 \\ (r_{X,Y})_k = \dfrac{n a_k - b_k d_k}{\sqrt{n f_k - b_k^2}\sqrt{n g_k - d_k^2}} \end{cases} \tag{5.18}$$

where a_k is the cumulative term of the product of two variables, b_k and d_k are the cumulative terms of variables X and Y, respectively, and f_k and g_k are the cumulative terms of the power of variables X and Y, respectively.

As described above, a moving window is used to calculate the SoC correlation coefficient to improve the robustness. Therefore, Eq. (5.18) can be rewritten as follows:

$$\begin{cases} A_k = a_k - a_{k-L} \\ B_k = b_k - b_{k-L} \\ D_k = d_k - d_{k-L} \\ F_k = f_k - f_{k-L} \\ G_k = g_k - g_{k-L} \\ (r_{X,Y})_k = \dfrac{L A_k - B_k D_k}{\sqrt{L F_k - B_k^2}\sqrt{L G_k - D_k^2}} \end{cases} \quad (k \geq L) \tag{5.19}$$

where L is the size of the moving window.

Noted that if L is significantly large, the detection sensitivity would be reduced. Therefore, an appropriate value of L should be selected to achieve high detection accuracy and short detection time, and it is set to 600 in this study.

5.2.2.2 Experimental Set-Up and Process

The real ISC is concealed inside the battery, and it is generally difficult to trigger it quantitatively. Common ISC experiments are carried out through some simulation experiments. The ideal ISC equivalent experiment needs to meet the following requirements [51]: (a) It can simulate the thermal and electrical behaviours of the battery; (b) The ISC resistance, trigger form, and time are controllable; (c) The battery damage is consistent with the actual situation; (d) High repeatability. At present, there is no ideal method that can meet all the above requirements. The commonly used method is to simulate the ISC with the external parallel resistance of the battery. This method has high repeatability and controllability, but it fails to reflect the thermal characteristics of ISC. Since the proposed ISC detection method does not involve the thermal characteristics of the battery, it is reasonable for us to use an external parallel resistor to simulate ISC. The experimental set-up is shown in Fig. 5.18, in

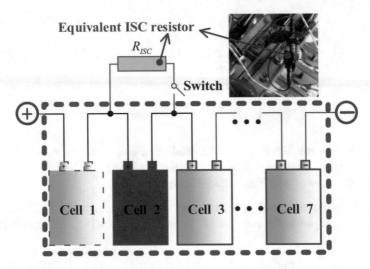

Fig. 5.18 Experimental device, reprinted from [48], with permission from Elsevier

which seven commercial ternary cells with the standard capacity of 50 Ah, cut-off voltage of 4.25 V, and discharge cut-off voltage of 2.8 V are connected in series for charging and discharging experiments. In addition, a resistor is connected in parallel to Cell 2 as the equivalent ISC resistance and controlled by a switch.

According to the Chinese National Standard GB/T 31484-2015, a normal battery should maintain 85% of its full capacity after being rest for 28 days in open circuit. Therefore, the critical ISC resistance can be roughly calculated as follows:

$$R_{\text{ISC}} = \frac{U_0}{C_0 * 15\%/(28 * 24)} = \frac{3.65}{50 \cdot 15\%/(28*24)} = 327\Omega \qquad (5.20)$$

where U_0 is the nominal voltage of the test cell; C_0 is the nominal capacity.

In the experiment, different ISC equivalent resistors (1, 10, and 100 Ω) are used to simulate different degrees of ISC. Specifically, the resistor of 100 Ω is used to simulate early slight ISC, resistor of 10 Ω is used to simulate developing ISC, and resistor of 1 Ω is used to simulate severe ISC. The charging and discharging schemes of this experiment can be described as follows. First, the experimental battery pack is charged to the cut-off voltage using a constant current, and then discharged under the New European Driving Cycle (NEDC) dynamic condition. The ISC control switch is closed at the same time, and the ISC resistor starts to work until the discharge cut-off voltage is reached. Repeat the above steps four times. Due to the extra power consumption in the external resistor, the voltage difference between the Cell 2 and the normal cell increases. Therefore, when using different ISC resistors to simulate different ISCs, the Cell 2 needs to be taken out from the module and charged it separately to ensure that its voltage is equivalent to the normal cell.

5.2.2.3 ISC Fault Diagnosis Results

Figure 5.19 shows the charge and discharge results of three equivalent ISC tests. It can be observed that the voltage of the Cell 2 is consistent with the initial voltage of other normal cells, indicating that the power loss caused by the ISC in the initial stage cannot be clearly distinguished. After the ISC is activated for a period, the voltage of Cell 2 is significantly lower than that of other normal cells, and the difference increases over time. In addition, increasing the severity of the ISC (i.e. reducing the ISC resistor) will produce an earlier voltage difference. As shown in Fig. 5.20b, when the cell is discharged to the cut-off voltage, the voltage difference between the Cell 2 and the normal cells reaches the maximum. This is because the open-circuit voltage

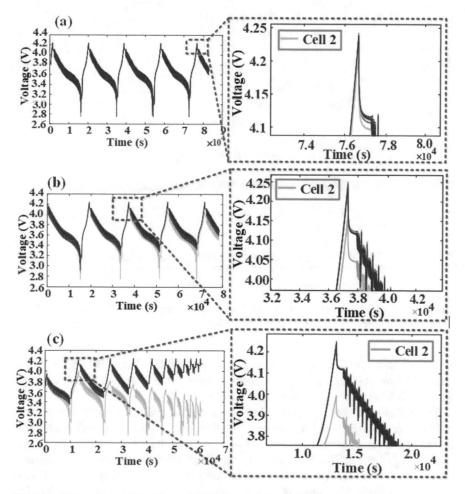

Fig. 5.19 Voltage curves for different ISC resistors under NEDC dynamic condition. **a** $R_{ISC} = 100\ \Omega$; **b** $R_{ISC} = 10\ \Omega$; **c** $R_{ISC} = 1\ \Omega$, reprinted from [48], with permission from Elsevier

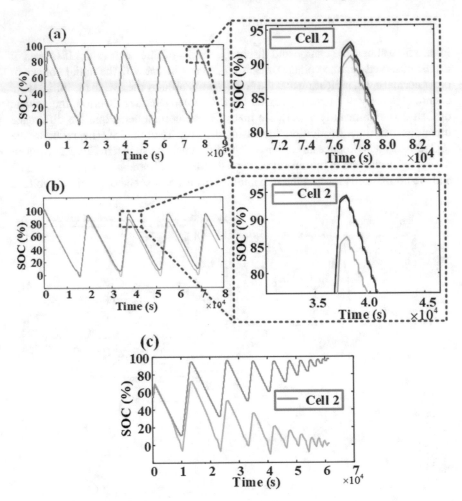

Fig. 5.20 SoC estimation results for different ISC resistors under the NEDC dynamic condition. **a** $R_{\text{ISC}} = 100\ \Omega$; **b** $R_{\text{ISC}} = 10\ \Omega$; **c** $R_{\text{ISC}} = 1\ \Omega$, reprinted from [48], with permission from Elsevier

of the cell at a low SoC shows the fastest downward trend, and the cell with ISC enters the low SoC region before the normal cells, thereby accelerating the voltage drop and further increasing the voltage difference.

In conclusion, these experimental results show that ISC can be detected and fault cells can be identified online through the difference of charging and discharging curves of the series cells. However, the voltage curve fluctuates greatly, which may lead to misjudgment. In addition, these voltage curves are obtained without considering sensor error and noise. Note that the voltage sensor error in the actual EVs is quite large, and the fluctuation of the actual voltage curve may be greater, which may affect the ISC detection accuracy. Therefore, it is not feasible to judge ISC directly through voltage curves.

ISC detection using SoC correlation coefficient. Based on the above experimental data and the EKF algorithm, the SoC of each cell is estimated, and then the SoC difference under different ISC resistors is studied. Figure 5.20 shows the SoC estimation results under NEDC dynamic conditions. As expected, the SoC curves are similar to the voltage curves described in Fig. 5.20, but they fluctuate much less than the voltage curves. Therefore, it is reasonable and feasible to detect ISC online by estimating the difference of SoC and setting a reasonable threshold, which can greatly reduce the false positive probability.

The SoC correlation coefficients of each cell and two adjacent cells are calculated, and the results are shown in Fig. 5.21. Here, $r_{i,j}$ is the SoC correlation coefficient between cells i and j, and the green dotted line indicates the ISC detection threshold. Only the two SoC correlation coefficients corresponding to a cell are simultaneously below the threshold, the cell can be judged as the ISC cell. To ensure the high accuracy and short detection time of ISC, the appropriate threshold is determined by offline calibration. From the experimental results, when the equivalent ISC resistance is 100 Ω and the SoC difference is higher than 2.5%, the correlation coefficient is 0.7, and thus, 0.7 is selected as the threshold in this study. Moreover, the initial SoC estimation at EV startup may have a large error to cause false alarms. Therefore, the car does not start the ISC detection until a period of time after it has been started. In this study, this time is set to 1000 s.

The enlarged graph on the right of Fig. 5.21 shows the SoC correlation coefficients between Cell 1 and Cell 2 (blue curve) and between Cell 2 and Cell 3 (red curve). In addition, when the two SoC correlation coefficients related to Cell 2 are lower than the threshold, an ISC alarm will be triggered, which is indicated by a red triangle. The SoC correlation coefficient of normal cells is above 0.8, indicating that the SoC between these cells is highly correlated. On the other hand, the SoC correlation coefficient of ISC cells with low SoC suddenly drops, indicating that the SoC correlation between the ISC cell and the normal cells is low. As shown in Fig. 5.21b, when the ISC resistance is less than 10 Ω, the SoC correlation coefficient is 0, indicating that the SoC of the ISC cell is completely different from that of the normal cells.

In general, it can be observed that the proposed online ISC detection algorithm is very sensitive and accurate under dynamic conditions. In addition, as shown in Fig. 5.21a, for the early ISC with a large equivalent resistance, although the detection time is longer, the method is still effective under dynamic conditions, and the detection time is significantly shorter than the latency of the early ISC. Therefore, the proposed method is effective for the online detection of early ISC.

Comparison with other ISC detection methods. The proposed data science-based ISC detection method is compared with the other three ISC detection methods to confirm its advantages, namely the static leakage, SoC difference, and voltage difference methods. The static leakage method is the simplest and most direct ISC detection method. According to GB/T31484-2015 [48], the battery should maintain 85% of its full capacity after being idle for 28 days in an open-circuit state. When the equivalent ISC resistance is 100, 10, and 1 Ω, the time required for the 15% capacity loss can

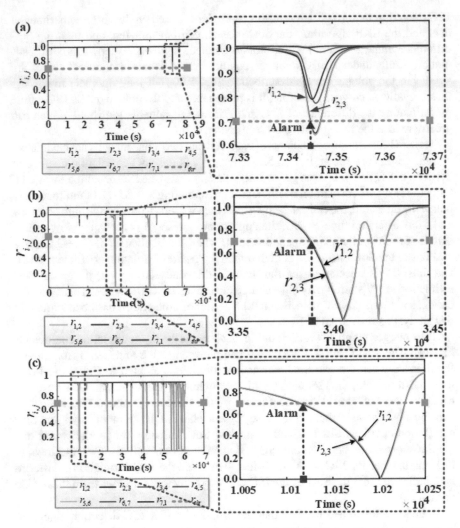

Fig. 5.21 ISC detection results based on the SoC correlation coefficient. **a** $R_{ISC} = 100\ \Omega$; **b** $R_{ISC} = 10\ \Omega$; **c** $R_{ISC} = 1\ \Omega$, reprinted from [48], with permission from Elsevier

be calculated as 202.7, 20.3, and 2.03 h according to Eq. (5.21), which is the ISC detection time.

$$t_d = \frac{C_0 \cdot 15\%}{28 \cdot 24 \cdot U_0 / R_{ISC}} \tag{5.21}$$

where t_d is the time required to detect the ISC.

The SoC difference method estimates the ISC based on the difference between the SoC of each cell and the average SoC. Since the cell with the smallest SoC is

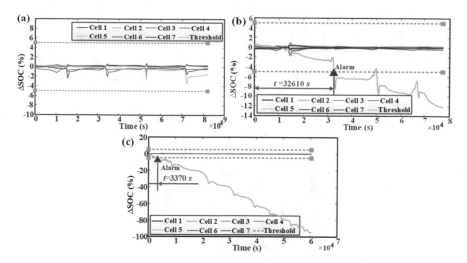

Fig. 5.22 ISC detection results using the SOC difference. **a** $R_{ISC} = 100\ \Omega$; **b** $R_{ISC} = 10\ \Omega$; **c** $R_{ISC} = 1\ \Omega$, reprinted from [48], with permission from Elsevier

more likely to have an ISC failure, this cell is not considered when calculating the average value:

$$\Delta SOC_i = SOC_i - \left(\sum_{i=1}^{7} SOC_i - \min\{SOC_1, \ldots SOC_7\}\right)/6 \qquad (5.22)$$

where ΔSOC_i is the difference between SoC of the ith cell and the average SoC.

In this study, a 5% SoC difference is set as the threshold for ISC detection; that is, if the cell's SoC difference is higher than 5%, it is considered that the cell has an ISC failure. Figure 5.22a shows that the SoC difference of the Cell 2 with an ISC resistance of 100 Ω increases very slowly and does not reach the threshold after 24 h, indicating that this method is not suitable for early ISC detection and the detection time is very long. Nevertheless, as the degree of ISC increases, the SoC difference increases significantly over time, thereby reducing the detection time.

Figure 5.23 shows the ISC detection results based on the voltage difference method. The threshold of voltage difference is set to 0.5 V in this study. It can be observed that the detection time using the voltage difference method for the early ISC is very long, and the voltage difference fluctuates greatly. Therefore, the voltage difference method is not suitable to detect the early ISC. It can be concluded that the proposed SoC correlation coefficient method has better detection accuracy and shorter detection time than the other three methods.

The battery has rich charge and discharge data in the service process, and the ISC will cause the change of battery voltage, which makes it possible to detect the ISC online and in real time based on the charge and discharge data. To improve the

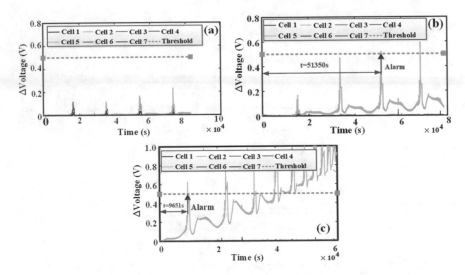

Fig. 5.23 ISC detection results using the voltage difference. **a** $R_{ISC} = 100\ \Omega$; **b** $R_{ISC} = 10\ \Omega$; **c** $R_{ISC} = 1\ \Omega$, reprinted from [48], with permission from Elsevier

reliability of the ISC detection, a data-driven ISC detection method based on SoC correlation coefficient is proposed in this section. The experimental results show that the proposed data science-based solution has the advantages of good real time, high accuracy and excellent robustness. It can detect the early ISC and greatly improve battery safety.

5.3 Battery Charging

Battery charging is also a key part required to be managed during battery operation [52]. Technical issues facing the development of efficient battery charging solutions arise from different charging objectives, hard constraints, and charging termination, as illustrated in Fig. 5.24. In this context, this section first introduces several key objectives that need to be considered during battery charging, then two case studies through designing suitable data science-based solutions for both battery cell charging and pack charging are detailed and analysed.

5.3.1 Battery Charging Objective

In general, the objective of designing suitable charging solution for Li-ion batteries is to provide a good capacity utilization, a short time for charging process, a high charging efficiency with less energy loss, while maintaining a long battery cycle

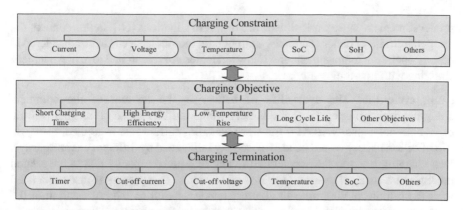

Fig. 5.24 Structures of charging infrastructure, objective as well as termination

life [53]. In addition, battery temperature will rise dramatically during the charging process especially in high power applications. Overheat temperature would result in battery failures so the temperature rise is also a critical objective for battery charging. Therefore, suitable battery operation management that provides the proper charging patterns to balance these charging objectives is indispensable for battery applications.

Short charging time: Charging time is one of the key aspects for the battery applications especially in transportation such as EVs. On the one hand, a long charging time will inevitably affect the convenience of EV usage and limit its acceptance by customers [54]. It is necessary to improve charging speed for EVs especially in some public charging conditions that are similar to gasoline refuelling for conventional vehicles. However, too fast charging will lead to significant energy loss and battery performance degradation, further decreasing battery performance or causing safety problems [55]. It is therefore rational to consider charging time as one of the key factors in designing battery charging solutions.

High energy efficiency: Battery energy efficiency is the ratio of the charged energy to the energy required to be discharged to the initial state prior to charging [56]. The energy efficiency during the battery charging process would be affected by many factors such as current, internal resistance, SoC, and temperature. Large energy loss implies low efficiency of energy conversion in battery charging, which needs to be addressed [57]. It is critical to develope optimal charging strategy that can decrease the energy loss caused by battery internal resistance and control charging or discharging currents appropriately to achieve high energy efficiency during the battery charging process.

Low-temperature rise: Temperature affects battery performance in many ways such as round trip efficiency, energy and power capability, cycle life, reliability, and charge acceptance [58]. Both the battery surface and internal temperatures may exceed permissible levels when it is charged with high current especially in high power applications [59], and the overheating temperatures may intensify battery ageing process and even cause explosion or fire in severe situations [60]. In this context, the temperature rise of battery becomes an important factor that needs to be considered in battery operation management and many strategies are researched to achieve battery charging with low-temperature rise.

Long cycle life: The cycle life of battery is the amount of the complete charging/discharging cycles that a battery works until its capacity falls below 80% of its nominal capacity. Even if the same charging current rate is applied, the cycle life of Li-ion battery would be also extremely influenced by different charging ways. Fast charging leads to the accelerated fading of battery capacity due to the related increase of a surface layer and the loss of Li+ ions. This process is also associated with the lithium plating onto the battery anode as well as the polarization at the electrode–electrolyte surface. According to the analysis of battery electrochemistry, a suitable charging current profile plays the important role in prolonging battery service life and needs to be carefully considered.

Other objectives: Some other objectives are also crucial for achieving efficient battery charging. Battery polarization, which means the variation of the equilibrium potential in a battery electrochemical reaction, has a tremendous impact on the battery charging performance. The battery charging current induces losses due to its polarization. Both the battery charging speed and efficiency would be enhanced by controlling battery polarization. In addition, decreased polarization contributes to the reduction of battery capacity fade, because the temperature rising rate can be restricted. So the charging polarization is also selected as an important objective to achieve battery health-conscious charging [61]. Some researchers also focus on developing charging strategies to increase battery available capacity [62], which could be achieved by the smaller increase of internal resistance or lower temperature rise.

Conflicting objectives: Developing an advanced charging strategy for battery operation management is not a simple task and usually implicates the trade-off among different coupled but conflicting objectives such as charging time and energy efficiency [63], while also pursuing temperature rise minimization, energy loss minimization, long service life, and low normalized discharged capacity. Therefore, it is essential to involve the optimization of multiple conflicting objectives when evaluating the cost-effectiveness of battery charging patterns in real battery charging applications.

The typical combinations of multi-objective for Li-ion battery charging can be divided into some main parts. For multi-stage constant-current charging, the double-objective optimization is often considered which combines the battery charging time and the normalized discharged capacity. Besides, the temperature rise, efficiency, and cycle life are also selected as the charging objectives in some published works [64, 65]. For the constant-current–constant-voltage charging, the main multi-objectives are often composed of battery charging time and energy efficiency. Besides, the temperature rise, cycle life, and capacity fading are also noticeable points for developing optimal charging strategies. For other charging strategies in published works, the same multi-objectives can be adopted as the optimization targets for improving battery charging performance during its operation management.

5.3.2   Case 1: Li-Ion Battery Economic-Conscious Charging

In this study, a data science-based framework through using multi-objective optimization solutions for economy-conscious charging is introduced, as shown in Fig. 5.25 [66]. Given the predefined battery electrothermal-ageing model and the economic price model from [67, 68], three important charging objective functions including battery charging time, average temperature, and particularly charging cost can be created. Then the suitable charging pattern is designed to charge battery with effective energy and time management.

5.3.2.1   MCC Profile

In this research, the typical multi-stage constant-current (MCC) charging pattern is explored. This MCC charging pattern generally consists of some CC phases, as illustrated in Fig. 5.26. Due to Li-ion battery being usually less susceptible to lithium plating at low SoC conditions, a relatively large CC rate I_{CC1} could be utilized at the beginning of MCC charging process to transfer enough energy throughput into a battery. Then a series of stepwise reduced CC phases would be adopted until reaching the last CC phase with a CC rate I_{CCN}. During this process, a CC phase would turn

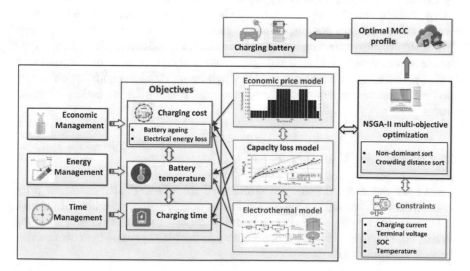

Fig. 5.25 Overall multi-objective optimization framework for economic-conscious charging, reprinted from [66], with permission from Elsevier

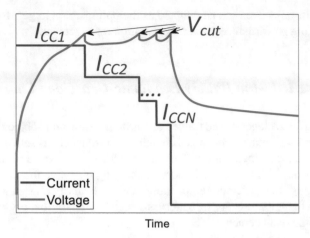

Time

Fig. 5.26 Charging current and terminal voltage for MCC profile, reprinted from [66], with permission from Elsevier

into another CC phase when the battery terminal voltage rises up to the predefined cut-off voltage V_{cut}.

Although this MCC pattern is convenient to be applied in EV applications, a key but challenging issue is to set suitable battery current and voltage values during its entire charging process. In theory, MCC pattern's charging time is mainly affected by battery cut-off voltage V_{cut} and current rates in each CC phase (I_{CC1}, I_{CC2},..., I_{CCN}). As recommended by Li-ion battery manufacturers, V_{cut} is generally set as its maximum level to improve battery capacity utilization. For charging current rate, a large value could directly speed up the charging process, but also lead to severe issues such as lithium plating, increased energy loss, and overheating of battery, further significantly affecting battery service life. Then the economic cost caused by the wasted electricity and faded battery capacity will be increased. In this context, it is vital to optimizing current rates of MCC pattern for efficient equilibration among the time, temperature, and particularly economic cost during the battery charging process.

5.3.2.2   Charging Cost Function

Based upon the coupled battery electrothermal model and related economic price model from [66], several crucial but conflicting cost functions could be formulated.

For battery charging time (BCT), less BCT represents that the battery charging process could be finished with a faster speed. In the MCC pattern, the cost function for BCT is expressed as:

$$\text{JMCC}_{\text{BCT}} = \Delta t * t\text{CC}_N / 60 \tag{5.23}$$

where Δt represents the sampling time period, tCC_N denotes the total amount of sample points when a battery is charged from its initial SOC to its final target SOC Here, JMCC$_{BCT}$ has an unit of minute (M).

For battery average temperature (BAT), less BAT could protect battery from overheating under same ambient temperature. The cost function JMCC$_{BAT}$ for BAT during MCC pattern is described as:

$$\text{JMCC}_{\text{BAT}} = \Delta t * \left[\sum_{t=0}^{tCC_1} T_a(t) + \cdots + \sum_{tCC_{N-1}+1}^{tCC_N} T_a(t) \right] / tCC_N \qquad (5.24)$$

where $tCC_1, \ldots, tCC_N$ are sample points when each CC phase is ended, respectively, $T_a(t)$ represents a radial average battery temperature at time point t.

For battery charging cost (BCC) that is divided into two main types including battery electricity loss cost (BEC) and battery ageing cost (BAC), its objective function JMCC$_{BCC}$ during MCC pattern can be described by:

$$\text{JMCC}_{\text{BCC}} = \text{JMCC}_{\text{BEC}} + \text{JMCC}_{\text{BAC}} \qquad (5.25)$$

where JMCC$_{BEC}$ denotes the charging cost caused by electricity loss and could be further described by:

$$\begin{cases} \text{JMCC}_{\text{BEC}} = \Delta t * \left[\sum_{t=0}^{tCC_1} f_{EC}(t) + \cdots + \sum_{tCC_{N-1}+1}^{tCC_N} f_{EC}(t) \right] \\ f_E(t) = a(t) * \left[I^2(t)R_0(t) + V_1^2(t)/R_1(t) + V_2^2(t)/R_2(t) \right] \end{cases} \qquad (5.26)$$

where $f_{EC}(t)$ represents electrical energy loss cost occurred at t, $a(t)$ denotes the corresponding TOU price at t. It should be known that the value of $a(t)$ remains constant for a long time period and would be affected by the time instant T of a day with a relation as follows:

$$a(t) = f(t \in T) = \begin{cases} \frac{1.1946}{3.6 \times 10^6} & T \in (23:00 - 7:00) \\ & T \in (7:00 - 10:00) \text{ or} \\ \frac{1.4950}{3.6 \times 10^6} & T \in (15:00 - 18:00) \text{ or} \\ & T \in (21:00 - 23:00) \\ \frac{1.8044}{3.6 \times 10^6} & T \in (10:00 - 15:00) \text{ or} \\ & T \in (18:00 - 21:00). \end{cases} \qquad (5.27)$$

For battery degradation cost during MCC charging, its cost function JMCC$_{BAC}$ could be obtained as:

$$\begin{cases} \text{JMCC}_{\text{BAC}} = \frac{B_{\text{new}} - B_{\text{used}}}{N_{\text{BAC}}} \\ N_{\text{BAC}} = \frac{T_{\text{Ah}}}{E_{\text{Ah}}} \end{cases} \qquad (5.28)$$

where B_{used} stands for the cost value of utilized battery, N_{BAC} represents the total amount of charging cycle when a battery's capacity degrades to its end-of-life (EoL). In addition, T_{Ah} is the total throughput in Ah for all charging cycles while E_{Ah} is that of one cycle. For EV applications, Li-ion battery's EoL is generally achieved when the capacity of battery cell degrades to 80% of its nominal value. Supposing that a battery is charged by using the same MCC pattern during each charging cycle, T_{Ah} would be mainly affected by the average $\tilde{I}$, $\widetilde{\text{soc}}$, and $\tilde{T}_a$ of this specific MCC pattern. Then T_{Ah} could be obtained by:

$$T_{\text{Ah}} = \left[\frac{20}{\sigma_{\text{funct}}\left(\tilde{I}, \widetilde{\text{soc}}, \tilde{T}_a\right)} \right]^{1/z} \tag{5.29}$$

$$\begin{cases} \tilde{I} = \sum_{t=0}^{tCC_N} I(t)/tCC_N \\ \widetilde{\text{soc}} = \sum_{t=0}^{tCC_N} \text{soc}(t)/tCC_N \\ \tilde{T}_a = \sum_{t=0}^{tCC_N} T_a(t)/tCC_N. \end{cases} \tag{5.30}$$

For the E_{Ah} of a specific MCC pattern, its charging throughput is also obtained as:

$$E_{\text{Ah}} = \frac{\Delta t}{3600} \sum_{t=0}^{tCC_N} I(t). \tag{5.31}$$

In this study, once the current rates of each CC phase are optimized, all elements of the cost functions JMCC_{BCT}, JMCC_{BAT}, and JMCC_{BCC} could be obtained. Namely, current rates I_{CC1}, I_{CC2},..., I_{CCN} become the decisive variables that require to be carefully optimized for battery economy-conscious charging.

5.3.2.3   Optimization Procedure

The optimization of MCC charging pattern belongs to a highly nonlinear and strongly coupled process. Numerous time-varied parameters such as capacitors and resistances within coupled battery model are strongly associated with battery temperatures and SoCs. Battery capacity degradation would be also highly affected by its electrical and thermal dynamics. Moreover, different constraints including the charging current, terminal voltage, battery SoC, and temperature need to be considered during optimization as follows:

$$\begin{cases} I_{\min} \leq I(t) \leq I_{\max} \\ V_{min} \leq V(t) \leq V_{\max} \\ s_I \leq \mathrm{SoC}(t) \leq s_T \\ T_{\min} \leq T_a(t) \leq T_{\max}. \end{cases} \tag{5.32}$$

For this time-varied and complex optimization problem, an effective multi-objective optimization tool is required to optimize MCC pattern and equilibrate these key but conflicting charging objectives. Due to the advantages of being immune from the NP-hard and highly nonlinear issues, meta-heuristic optimization method becomes a powerful tool. Among different meta-heuristic multi-objective optimizers, non-dominated sorting genetic algorithm II (NSGA-II) has been widely used in many real applications for handling complicated optimization issues due to its outstanding ability to keep elitism optimization and preserve diversity. In the light of this, NSGA-II approach is herein employed to search the optimal MCC patterns, with the purpose of achieving economy-conscious charging under various priorities.

Figure 5.27 illustrates the optimization flowchart of searching the proper MCC pattern to equilibrate battery charging time, average temperature, and particularly economic cost. Each process is detailed as follows:

Step 1: Set battery charging constraints. These constraints include: (1) the initial and target SoCs:s_I and s_T; (2) the current, voltage, and temperature limitations: $I_{\min}$, $I_{\max}$, $V_{\min}$, $V_{\max}$, $T_{\min}$, and $T_{\max}$.

Step 2: Set MCC pattern's parameters. These parameters include: (1) the cut-off voltage: V_{cut}; (2) the number of CC charging phases: N_{CC}.

Step 3: Select charging objectives with the consideration of different user requirements. According to the priority of battery application, formulating a proper cost function with a combination of BCT, BAT, and BCC.

Step 4: Initialize NSGA-II's parameters. The main parameters of NSGA-II include: (1) the number of generation: G_m; (2) the population size: N_p.

Step 5: For $k = 1$ to $k_{\max}$, do

(1) For each CC phase, calculating the cost function combining the selected charging objectives until battery SoC reaches a target s_T. When battery terminal voltage goes up to its V_{cut}, a CC phase would be terminated and then jump into another one.

(2) Evaluate $\mathrm{JMCC_{BCT}}$, $\mathrm{JMCC_{BAT}}$, and $\mathrm{JMCC_{BCC}}$ of whole charging process through summing the cost functions of each CC phase. Then analyse the optimized MCC pattern's performance based on the optimal sets from related Pareto frontier.

(3) Search a new MCC pattern again if the obtained MCC pattern is unsatisfactory. This optimization process will be terminated when an end criterion $k_{\max}$ is reached.

Through following this data science framework, economy-conscious battery charging could be achieved based on the optimized MCC pattern. Moreover, battery charging time and average temperature could be also balanced.

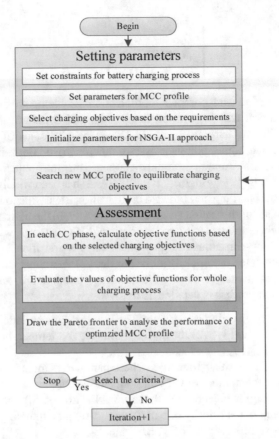

Fig. 5.27 Optimization flowchart by searching the proper MCC profile to equilibrate charging objectives, reprinted from [66], with permission from Elsevier

5.3.2.4 Results and Discussion

Then the sensitivity of key parameters including battery cut-off voltage, convection resistance, and ambient temperature is first analysed to explore their effects on the MCC pattern optimization. The optimal set is drawn by the form of Pareto frontier to equilibrate conflicting objectives for various user demands. Constant parameters are set as: $\Delta t = 1s$, $s_I = 0.1$, and $s_T = 0.95$, respectively. Temperatures are set as: $T_s(0) = T_I(0) = T_{amb}$. Battery parameter limitations are set as: $I_{min} = 0\,A$, $I_{max} = 10\,A$, $V_{min} = 3.0\,V$, $V_{max} = 3.6\,V$, $T_{min} = 15\,°C$, and $T_{max} = 45\,°C$. The population size and generation number of NSGA-II are set as $N_p = 360$ and $G_m = 60$.

Sensitivity Analysis for the MCC Profile

For the multi-objective optimizations, the Pareto frontier could provide a series of optimal strategies to graphically demonstrate cases that one cost function cannot

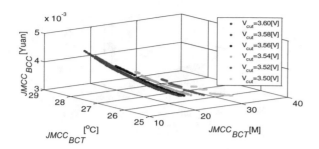

Fig. 5.28 Effects of cut-off voltages on the optimized MCC profile via Pareto frontier, reprinted from [66], with permission from Elsevier

be enhanced without making other cost functions worse. If the Pareto frontier from a specific case could distribute evenly and widely, while the points in this Pareto frontier are close to the origin of coordinate, the optimal MCC pattern for such case becomes better. In the light of this, sensitivities of cut-off voltage, convection resistance, and ambient temperature are analysed via the Pareto frontier to explore their influences on battery charging.

Sensitivity of cut-off voltage: in theory, a small cut-off voltage V_{cut} of battery would restrain battery capacity utilization, but too large V_{cut} can also cause damages or safety issues to battery. In this test, I_{CC1} is initially set within 5A (2C) and 10A (4C). The convection resistance R_u and ambient temperature T_{amb} are set as 3.08 KW^{-1} and 25 °C, respectively. Six V_{cut} cases (3.60, 3.58, 3.56, 3.54, 3.52, and 3.50 V) are utilized to explore battery cut-off voltage's sensitivity.

Figure 5.28 illustrates the Pareto frontiers for the optimized MCC patterns with various V_{cut}. Obviously, a better optimal MCC set could be obtained through using a larger V_{cut}. As V_{cut} reduces, the number of obtained optimal points would also decrease, and the Pareto frontier tends to move to the right, leading to a fact that more time is spent for battery charging. In addition, the values of both JMCC$_{BAT}$ and JMCC$_{BCC}$ would become less with the reduced V_{cut}. It can be concluded that a small V_{cut} leads to the low charging current, and hence, the average temperature rise of battery is also restrained, further causing less cost of charging, but charging speed would be sacrificed accordingly.

Sensitivity of convection resistance: another key parameter is the heat convection resistance R_u that reflects heat convection between ambient condition and battery shell. This parameter could also reflect battery thermal management operation such as the air fan or liquid cooling. For this test, I_{CC1} is also set within 5A and 10A. V_{cut} is fixed as 3.6 V while T_{amb} is set as 25 °C. Five cases of R_u (1, 3.08, 5, 10, and 15 KW^{-1}) are utilized to explore R_u's sensitivity.

After the optimization, the Pareto frontiers for optimized MCC pattern under different R_u are illustrated in Fig. 5.29. It can be seen that as R_u increases, the Pareto frontier would move to the upper. On the contrary, battery charging process would

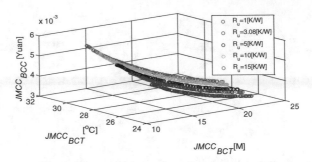

Fig. 5.29 Effects of convection resistances on the optimized MCC profile via Pareto frontier, reprinted from [66], with permission from Elsevier

speed up when increasing R_u. To summarize, a lower R_u favours battery average temperature as well as the charging cost, at a sacrifice of charging time.

Sensitivity of ambient temperature: for the test of ambient temperature T_{amb}, I_{CC1} is initially set between 5 and 10A. V_{cut} and R_u are set as 3.6 V and 3.08 KW^{-1}, respectively. Six T_{amb} cases (15 °C, 20 °C, 25 °C, 30 °C, 35 °C, and 40 °C) are utilized to explore T_{amb}'s sensitivity.

Figure 5.30 illustrates the influences of T_{amb} on the optimized MCC pattern. Obviously, larger T_{amb} will lead the obtained Pareto frontier move to the left-upper, indicating that charging time is reduced but both the battery average temperature and charging cost would become larger adversely. Therefore, T_{amb} with small value favours achieving low charging cost as well as average temperature rise, but it would also lead battery internal resistance become larger, further inducing a reduced charging speed of battery.

In summary, the optimized MCC charging pattern is critically associated with these three parameters. Large V_{cut} benefits battery charging time. Low R_u and small T_{amb} favour the charging cost and average temperature of battery. It is vital to carefully select suitable parameters, based on specific requirements from various battery applications.

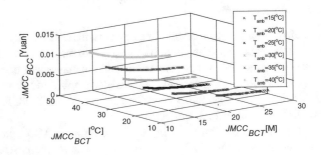

Fig. 5.30 Effects of the ambient temperature on the optimized MCC profile via Pareto frontier, reprinted from [66], with permission from Elsevier

Tests for Various Charging Cases

Next, the optimization results using different two charging objective combinations are first analysed, followed by several balanced MCC charging patterns to explore their efficacies on the trade-offs of various charging objectives. In this test, I_{CC1} is initially fixed within 5 and 10 A. The current decrement of CC phases is set between 1.5 and 2.5 A. V_{cut} is set as 3.6 V, while R_u and T_{amb} are set as $3.08 KW^{-1}$ and 25°C, respectively. After using NSGA-II to optimize MCC pattern, the corresponding Pareto frontiers of various two charging objective combinations are shown in Fig. 5.31.

From Fig. 5.31a, b, the optimal particles in the Pareto frontiers could be distributed uniformly in a wide region, indicating that both charging cost and battery average temperature conflict with charging speed. Not surprisingly, a quicker charging speed can be obtained through adopting a larger current rate in the MCC pattern. However, this increased charging current would also result in the larger thermal reactions happening within a battery, hence increasing the battery average temperature. On the other hand, both the average SoC and current during the whole charging process would be higher under a larger current rate case. This would aggravate the electrical energy loss and battery fading during charging. In this context, related economic charging costs will be intensified. From Fig. 5.31c, it can be seen that the optimal particles finally converge into a single point, indicating that there is no conflict between charging cost and battery average temperature.

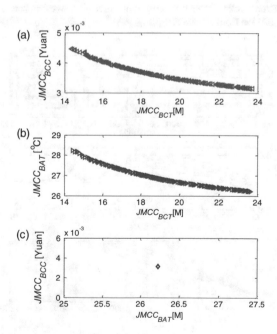

Fig. 5.31 Optimization results of the MCC profile with two objectives via Pareto frontier, reprinted from [66], with permission from Elsevier

Balanced charging: Fig. 5.32 illustrates the optimization results of MCC patterns with three balanced objectives. It can be seen that the objective functions JMCC$_{BAT}$ and JMCC$_{BCC}$ conflict with JMCC$_{BCT}$. Through sacrificing charging time, both battery temperature rise and economic loss could be substantially restrained. According to five different cases of balanced charging, Case 1 and Case 2 represent the quick charging solutions that spend less time to reach SoC target. Case 4 and Case 5 prefer lower charging cost and average temperature rise. Case 3 denotes a neutral position. Then Fig. 5.33 illustrates electrical dynamics while Fig. 5.34 shows the thermal dynamics of these balanced charging cases, respectively.

According to Figs. 5.33 and 5.34, as the required charging time becomes larger from Case 1 to Case 5, the initial current in the first CC phase inevitably decreases.

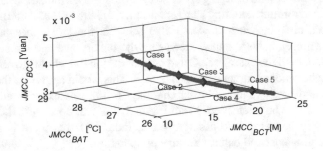

Fig. 5.32 Optimization results of the MCC profile with three objectives via Pareto frontier, reprinted from [66], with permission from Elsevier

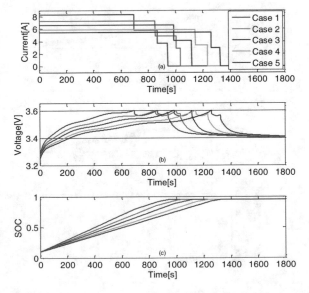

Fig. 5.33 Electrical dynamics of the MCC profile for the balanced charging: **a** current, **b** voltage, and **c** SoC, reprinted from [66], with permission from Elsevier

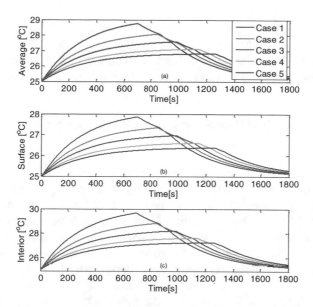

Fig. 5.34 Thermal dynamics of the MCC profile for the balanced charging: **a** average temperature, (**b** surface temperature, and **c** internal temperature, reprinted from [66], with permission from Elsevier

The total CC phase number within MCC pattern is also reduced gradually. Quantitatively, compared with Case 1 owns the quick charging solution that takes 15.72 min, Case 5 spends 22.10 min charging time (41% increase). However, battery average temperature and total charging cost in such a case, respectively, decrease to 26.41 °C (4.5% decrease) and 0.0033 *Yuan* (19.5% decrease). Through shortening charging time, both the average SoC and temperature of battery will increase dramatically to accelerate electricity loss and battery degradation, further leading to the increased economic cost during charging. Moreover, there is a range that increasing charging time linearly would decrease charging cost and average battery temperature significantly. Outside this range, further linearly speeding up the charging process would cause less effect to reduce charging cost and battery temperature. From this test, MCC patterns before Case 3 could be adopted to effectively decrease both economic cost and average temperature rise.

5.3.3 Case 2: Li-Ion Battery Pack Charging with Distributed Average Tracking

In this study, a data science-based charging solution for Li-ion battery pack is explored with low computational burden [69]. To be specific, based upon the typical Rint models for battery cells, an optimized average SoC trajectory could be first generated by constructing and handling a multi-objective optimization considering both user demand and the energy loss of battery pack. Then a distributed charging solution is derived to make the SoCs from cells could track this trajectory, where the model bias observers of each cell are designed for online compensation.

5.3.3.1   User-Involved Data Science Charging Framework for Battery Pack

Figure 5.35 illustrates the explored multi-module charger for Li-ion battery pack with n cells connected in serial. For this multi-module charge, each cell could be charged by an independent module, bringing the benefits to avoid the overcharging issue. Due to the superiorities in terms of easy to be implemented, reasonable costs and size, this type of charger has been successfully adopted in many real applications.

According to the concept of leader–followers in the multi-agent system [70], a distributed average tracking framework can be designed to charge each cell of battery pack, as illustrated in Fig. 5.36. Specifically, an average charging trajectory is first derived as the leader role. Then all cells as the follower role could track this trajectory simultaneously. To achieve this, a typical cell Rint model with an average initial SoC is adopted to produce an optimized average charging trajectory firstly as:

$$\begin{cases} x_0(k+1) = x_0(k) + b_0 u_0(k) \\ y_0(k) = f_0(x_0(k)) + h_0(x_0(k))u_0(k) \end{cases} \tag{5.33}$$

where x_0, y_0, and u_0 represent Rint model's state, output, and input, respectively. b_0, $f_0(\cdot)$, and $h_0(\cdot)$ are related nominal values. Here b_i is a Coulomb efficiency parameter, $f_i(\cdot)$ and $h_i(\cdot)$ ($1 \leq i \leq n$) stand for the nonlinear relations between SoC and OCV as well as internal resistance, respectively. The initial value $x_0(0)$ is the average value of all cells' initial SoC. Then the optimal average SoC trajectory could be obtained by handling an optimization issue as:

$$\min_{u_0(0),\dots,u_0(N-1)} \gamma_1 (x_0(N) - x_s)^2 + \gamma_2 \sum_{k=0}^{N-1} h_0(x_0(k))u_0^2(k) \tag{5.34}$$

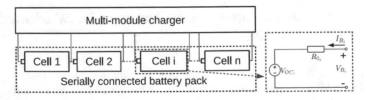

Fig. 5.35 Multi-module charger for battery packs, reprinted from [69], with permission from IEEE

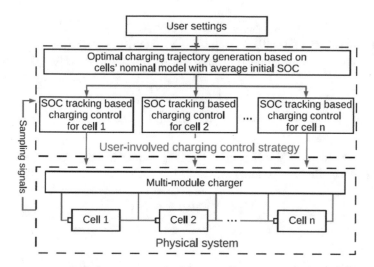

Fig. 5.36 Framework of derived user-involved data science charging solution, reprinted from [69], with permission from IEEE

$$\text{s.t. } x_0(k+1) = x_0(k) + b_0 u_0(k)$$

$$f_0(x_0(k+1)) + h_0(x_0(k+1))u_0(k) \leq y_M$$

$$0 \leq u_0(k) \leq u_M, \ x_0(k+1) \leq x_M \tag{5.35}$$

Supposing $U = [u_0(0), \ldots, u_0(N-1)]^T$, $H_k = \left[1_k^T, 0_{N-k}^T\right]$, Eq. (5.34) can be expressed as $x_0(k) = x_0(0) + b_0 H_k U$ and the related optimization issue could be further rewritten as:

$$\begin{cases} \underset{U}{\min} \ J_1(U) \\ \text{s.t. } C(U) \leq 0 \end{cases} \tag{5.36}$$

with

$$J_1(U) = U^T \left(\gamma_1 b_0^2 H_N^T H_N + \gamma_2 G(U)\right) U \\ + 2\gamma_1 b_0(x_0(0) - x_s) H_N U + \gamma_1(x_0(0) - x_s)^2 \tag{5.37}$$

$$C(U) = \begin{bmatrix} F(U) + G_1(U)U - Y_M \\ \Phi U - U_M \\ M U - X_C \end{bmatrix} \tag{5.38}$$

where U represents the optimization variable. Other variables are expressed as:

$$\begin{cases} G(U) = \text{diag}\{h_0(x_0(0)), \ldots, h_0(x_0(0) + b_0 H_{N-1} U)\} \\ F(U) = [f_0(x_0(0) + b_0 H_1 U), \ldots, f_0(x_0(0) + b_0 H_1 U)]^T \\ G_1(U) = \text{diag}\{h_0(x_0(0) + b_0 H_1 U), \ldots, h_0(x_0(0) + b_0 H_N U)\} \\ Y_M = y_M 1_N \\ \Phi = \begin{bmatrix} I_N, & -I_N \end{bmatrix}^T \\ U_M = \begin{bmatrix} u_M 1_N^T, & 0_N^T \end{bmatrix}^T \end{cases} \qquad (5.39)$$

where I_N is an identity $N \times N$ matrix. Then the optimal average SoC trajectory $x_0^r(k)$ ($1 \leq k \leq N$) could be scheduled as:

$$x_0^r(k) = x_0(0) + b_0 H_k U^r \qquad (5.40)$$

5.3.3.2   Distributed Battery Charging Based on SoC Tracking

For the ith ($1 \leq i \leq n$) cell, a SoC-tracking-based solution is derived to charge battery, as shown in Fig. 5.37, where a model bias observer is designed as:

$$\begin{cases} \widehat{w}_i(k+1) = \widehat{w}_i(k) + l_i\big(y_i(k) - \widehat{y}_i(k)\big) \\ \widehat{y}_i(k) = f_0(x_i(k)) + h_0(x_i(k))u_i(k) + \widehat{w}_i(k) \end{cases} \qquad (5.41)$$

where $\widehat{w}_i(k)$ and l_i represent estimated result and the observer gain, respectively.

To ensure the SoC $x_i(k)$ of ith cell could track the obtained trajectory $x_r^0(k)$ while also meet the constraint requirements, the SoC-tracking-based solution can be formulated as:

$$\min_{u_i(k)} \frac{\gamma_3}{2}\big(x_i(k) - x_0^r(k)\big)^2 + \frac{\gamma_4}{2}\gamma_2 u_i^2(k) \qquad (5.42)$$

$$\text{s.t. } x_i(k+1) = x_i(k) + b_i u_i(k)$$
$$f_0(x_i(k+1)) + h_0(x_i(k+1))u_i(k) + \widehat{w}_i(k+1) \leq y_M \qquad (5.43)$$
$$0 \leq u_i(k) \leq u_M, \ x_i(k+1) \leq x_M$$

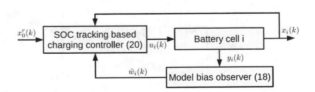

Fig. 5.37 Diagram of SoC-tracking-based charging solution for ith battery cell, reprinted from [69], with permission from IEEE

where γ_3 and γ_4 denote weight parameters that should be over zero. Then battery charging current $u_i(k)$ can be derived through handling the optimization issue with constraints at each k to make battery cells could track their optimized trajectories.

5.3.3.3 Results and Discussions

In this study, a battery pack with 10 cells connected in serial is utilized to explore the performance of designed data science-based charging solution. Table 5.9 gives the detailed information of cell's capacity and initial SoC. Nonlinear relations of SoC and OCV $f_0(\cdot)$ as well as SoC and internal resistance $h_0(\cdot)$ are described by:

$$f_0(\cdot) = \begin{cases} a_1 SoC + b_1, & \text{if } 0 \le SoC < 0.05 \\ a_2 SoC + b_2, & \text{if } 0.05 \le SoC < 0.20 \\ a_3 SoC + b_3, & \text{if } 0.20 \le SoC < 1.00 \end{cases} \quad (5.44)$$

$$h_0(\cdot) = \begin{cases} a_4 SoC + b_4, & if \ 0 \le SoC < 0.05 \\ b_5, & if \ 0.05 \le SoC < 0.95 \\ a_6 SoC + b_6, & if \ 0.95 \le SoC < 1.00 \end{cases} \quad (5.45)$$

where $a_1 = 3.61$, $b_1 = 3.13$, $a_2 = 1.21$, $b_2 = 3.2$, $a_3 = 0.8$, $b_3 = 3.282$, $a_4 = -0.46$, $b_4 = 0.057$, $b_5 = 0.034$, $a_6 = 2.06$, $b_6 = -1.923$. The maximum current and terminal voltage of each cell are 3C and 4.2 V. Here the weight coefficients are set as: $\gamma_1 = 10^4$, $\gamma_2 = 0.1$, $\gamma_3 = 10^4$, and $\gamma_4 = 10^{-3}$. Sample periods T_1 and T_2 are set as 300 s and 1 s, respectively.

Charging results: in this study, the target SoC x_s and charging time T_s are set as 100% and 120 min. Figure 5.38a–c illustrates the charging results of SoC, current and terminal voltage of each cell through using the designed data science strategy. It can be seen their constraints are all well guaranteed. According to the related battery pack SoC, cell SoC difference, and energy loss as shown in Fig. 5.38d–f, the SoC of battery pack could successfully track the scheduled average trajectory to reach the desired SoC of 98.48%. In addition, the SoC difference of cells would converge

Table 5.9 Capacities and initial SoC of different cells within a pack, reprinted from [69], with permission from IEEE

	Cell 1	Cell 2	Cell 3	Cell 4	Cell 5
Capacity (Ah)	10.04	9.95	9.96	10.05	9.97
Initial SoC (%)	8	6	9	12	7
	Cell 6	Cell 7	Cell 8	Cell 9	Cell 10
Capacity (Ah)	10.01	9.99	10	9.98	10.03Ah
Initial SoC (%)	10	11	3	5	4

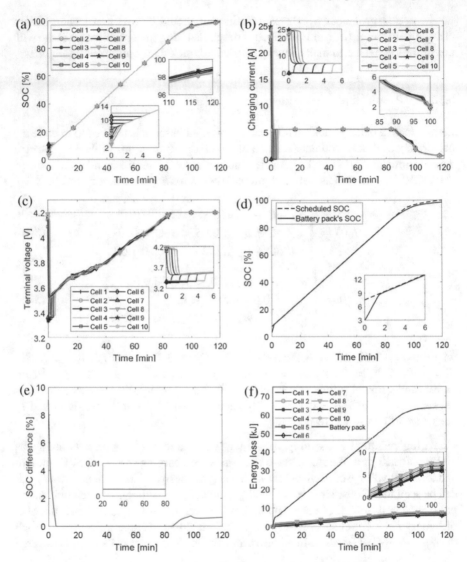

Fig. 5.38 Charging results for battery pack: **a** SoCs of battery cells, **b** terminal voltages of battery cells, **c** Charging currents of battery cells, **d** SoC of battery pack, (e) SoC difference of cells, **f** energy loss with $x_s = 100\%$ and $T_s = 120$ min, reprinted from [69], with permission from IEEE

from 9.08 to 0.64%, further validating the effectiveness of the designed charging strategy. The cells' SoC difference would become larger when the cell is nearly fully charged. This is not surprising as a constant-voltage stage is obtained at this region to meet voltage constraint, which would further lead to the inconsistent current to charge each cell.

Tests of various user settings: to further investigate the performance of the proposed charging strategy under various user settings, two tests in terms of various charging durations and desired SoCs are carried out. For test of various charging durations, x_s is set as 100%, charging duration T_s is set as 120 min, 90 min, 60 min, and 30 min, respectively. For test of desired SoC, T_s is set as 120 min, while x_s is set as 100%, 90%, and 80%, respectively. Based upon these settings, the corresponding SoC responses of battery pack for these two tests are shown in Fig. 5.39. Table 5.10 also illustrates the results of battery pack SoC, energy loss as well as SoC difference at the end of battery charging. Quantitatively, the SoC of battery pack can be rapidly charged to 77.76% with a relatively larger energy loss of 134.8 kJ through using a tight charging duration of 30 min. On the contrary, through using a long charging duration of 120 min, battery pack could be charged to 79.93% with a relatively lower energy loss of 36.5 kJ. This indicates another benefit of using the derived charging strategy, as the charging current could be tuned based on different user requirements to reduce the energy loss of pursuing rapid charging blindly.

Tests of various weight selections: the coefficients γ_1 and γ_2 reflect the weights of user demand as well as energy loss. To explore the effects of different weights on battery charging performance, γ_1 is fixed in this study while γ_2 is set as 0.01, 0.1,

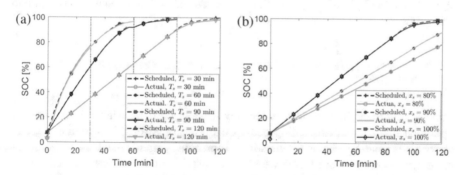

Fig. 5.39 SoC responses of battery pack: **a** various charging durations, **b** various desired battery SoCs, reprinted from [69], with permission from IEEE

Table 5.10 Results of the end of charging process under various user settings, reprinted from [69], with permission from IEEE

	Battery pack's SoC [%]	Energy loss [KJ]	SoC difference [%]
$x_s = 100\%, T_s = 30$ min	77.76	134.8	0.03
$x_s = 100\%, T_s = 60$ min	96.84	142.8	0.17
$x_s = 100\%, T_s = 90$ min	98.62	105.6	0.61
$x_s = 100\%, T_s = 120$ min	98.43	63.8	0.64
$x_s = 90\%, T_s = 120$ min	89.93	46.1	0.002
$x_s = 80\%, T_s = 120$ min	79.93	36.5	0.002

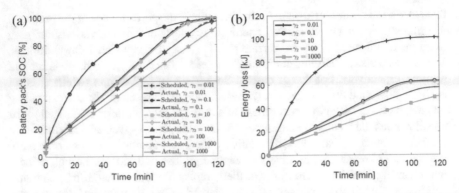

Fig. 5.40 Battery pack charging results: **a** battery pack's SoC, **b** energy loss under various weight coefficients, reprinted from [69], with permission from IEEE

10, 100, and 1000, respectively. Figure 5.40 shows the related results of battery pack SoC and energy loss. From Fig. 5.40, obviously increasing γ_2 could result in less energy loss, but would adversely affect the performance of charging battery pack to a predefined value. It can be seen that a proper trade-off could be achieved by setting γ_2 as 0.1. In the light of this, $\gamma_2 = 0.1$ is selected for designing charging solution in this study.

5.4 Summary

This chapter describes another three key aspects of data science-based battery operation management including battery ageing prognostics, fault diagnosis, and charging. For battery ageing prognostics, Li-ion battery ageing mechanism and stress factors are first introduced, followed by the description of the data science framework and classical methods to achieve battery ageing/lifetime prediction. Then two data science-based case studies through deriving modified GPR and a hybrid data science model to predict future cyclic capacity degradation and battery RUL are given and discussed. For battery fault diagnosis, after overviewing three typical types of data science-based methods, a data science-based case study of deriving a battery ISC fault detection strategy through using SoC correlation is described. For battery charging, several key and conflicting objectives during battery charging are first introduced, then two data science-based case studies through designing a multi-objective optimization-based battery cell economic-conscious charging and a distributed average tracking-based battery pack charging are introduced, respectively. All these case studies indicate that through designing suitable data science-based strategies, satisfactory results of battery ageing prognostics, fault diagnosis, and charging can be achieved for effective battery operation management.

References

1. Palacín MR (2018) Understanding ageing in Li-ion batteries: a chemical issue. Chem Soc Rev 47(13):4924–4933
2. Yu X, Manthiram A (2018) Electrode–electrolyte interfaces in lithium-based batteries. Energy Environ Sci 11(3):527–543
3. Barré A, Deguilhem B, Grolleau S, Gérard M, Suard F, Riu D (2013) A review on lithium-ion battery ageing mechanisms and estimations for automotive applications. J Power Sources 241:680–689
4. Lucu M, Martinez-Laserna E, Gandiaga I, Liu K, Camblong H, Widanage W, Marco J (2020) Data-driven nonparametric Li-ion battery ageing model aiming at learning from real operation data. Part A: Storage operation. J Energy Storage 30:101409
5. Lucu M, Martinez-Laserna E, Gandiaga I, Liu K, Camblong H, Widanage W, Marco J (2020) Data-driven nonparametric Li-ion battery ageing model aiming at learning from real operation data. Part B: Cycling operation. J Energy Storage 30:101410
6. Liu K, Li Y, Hu X, Lucu M, Widanage WD (2019) Gaussian process regression with automatic relevance determination kernel for calendar aging prediction of lithium-ion batteries. IEEE Trans Industr Inform 16(6):3767–3777
7. Li , K. Liu, A.M. Foley, A. Zülke, M. Berecibar, E. Nanini-Maury, J. Van Mierlo, H.E. Hoster, Data-driven health estimation and lifetime prediction of lithium-ion batteries: a review. Renew Sust Energ Rev 113:109254
8. Birkl CR, Roberts MR, Mcturk E, Bruce PG, Howey DA (2017) Degradation diagnostics for lithium ion cells. J Power Sources 341:373–386
9. Tang X, Liu K, Li K, Widanage WD, Kendrick E, Gao F (2021) Recovering large-scale battery aging dataset with machine learning. Patterns 2(8):100302 (2021)
10. Zhang F, Xiao L, Coskun D, Pang H, Xie S, Liu K, Cui Y (2022) Comparative study of energy management in parallel hybrid electric vehicles considering battery ageing. Energy 123219
11. De Hoog J, Jaguemont J, Nikolian A, Van Mierlo J, Van Den Bossche P, Omar N (2018) A combined thermo-electric resistance degradation model for nickel manganese cobalt oxide based lithium-ion cells. Appl Therm Eng 135:54–65
12. Schimpe M, Von Kuepach ME, Naumann M, Hesse HC, Smith K, Jossen A (2018) Comprehensive modeling of temperature-dependent degradation mechanisms in lithium iron phosphate batteries. J Electrochem Soc 165(2):A181
13. Su L, Zhang J, Huang J, Ge H, Li Z, Xie F, Liaw BY (2016) Path dependence of lithium ion cells aging under storage conditions. J Power Sources 315:35–46
14. Schmalstieg J, Käbitz S, Ecker M, Sauer DU (2014) A holistic aging model for Li (NiMnCo) O_2 based 18650 lithium-ion batteries. J Power Sources 257:325–334
15. Sarasketa-Zabala E, Gandiaga I, Rodriguez-Martinez L, Villarreal I (2014) Calendar ageing analysis of a LiFePO$_4$/graphite cell with dynamic model validations: Towards realistic lifetime predictions. J Power Sources 272:45–57
16. Hu C, Ye H, Jain G, Schmidt C (2018) Remaining useful life assessment of lithium-ion batteries in implantable medical devices. J Power Sources 375:118–130
17. Su X, Wang S, Pecht M, Zhao L, Ye Z (2017) Interacting multiple model particle filter for prognostics of lithium-ion batteries. Microelectron Reliab 70:59–69
18. Hu C, Jain G, Tamirisa P, Gorka T (2014) IEEE, method for estimating capacity and predicting remaining useful life of lithium-ion battery. In: proceedings of IEEE annual international conference on Prognostics and Health Management (PHM), Spokane, WA
19. Saha B, Goebel K (2009) Modeling Li-ion battery capacity depletion in a particle filtering framework. In: Proc Ann Conf PHM Soc 1(1)
20. Zhang L, Mu Z, Sun C (2018) Remaining useful life prediction for lithium-ion batteries based on exponential model and particle filter. IEEE Access 6:17729–17740
21. Xing Y, Ma EW, Tsui K-L, Pecht M (2013) An ensemble model for predicting the remaining useful performance of lithium-ion batteries. Microelectron Reliab 53(6):811–820

22. He W, Williard N, Osterman M, Pecht M (2011) Prognostics of lithium-ion batteries based on Dempster-Shafer theory and the Bayesian Monte Carlo method. J Power Sources 196(23):10314–10321

23. Chang Y, Fang H, Zhang Y (2017) A new hybrid method for the prediction of the remaining useful life of a lithium-ion battery. Appl Energy 206:1564–1578

24. Zhang H, Miao Q, Zhang X, Liu Z (2018) An improved unscented particle filter approach for lithium-ion battery remaining useful life prediction. Microelectron Reliab 81:288–298

25. Duong PLT, Raghavan N (2018) Heuristic Kalman optimized particle filter for remaining useful life prediction of lithium-ion battery. Microelectron Reliab 81:232–243

26. Ma Y, Chen Y, Zhou X, Chen H (2018) Remaining useful life prediction of lithium-ion battery based on Gauss-Hermite particle filter. IEEE Trans Control Syst Technol 27(4):1788–1795

27. Sun Y, Hao X, Pecht M, Zhou Y (2018) Remaining useful life prediction for lithium-ion batteries based on an integrated health indicator. Microelectron Reliab 88:1189–1194

28. Burgess WL (2009) Valve regulated lead acid battery float service life estimation using a Kalman filter. J Power Sources 191(1):16–21

29. Tang X, Liu K, Wang X, Liu B, Gao F, Widanage WD (2019) Real-time aging trajectory prediction using a base model-oriented gradient-correction particle filter for Lithium-ion batteries. J Power Sources 440:227118

30. Richardson RR, Osborne MA, Howey DA (2017) Gaussian process regression for forecasting battery state of health. J Power Sources 357:209–219

31. Hu X, Yang X, Feng F, Liu K, Lin X (2021) A particle filter and long short-term memory fusion technique for lithium-ion battery remaining useful life prediction. J Dyn Syst Meas Control Trans ASME 143(6):061001

32. Saha B, Goebel K, Poll SJ (2007) Christophersen, Ieee, An integrated approach to battery health monitoring using Bayesian regression and state estimation. In: Proceedings of 42nd annual AUTOTESTCON conference, Baltimore, MD, pp 646–653

33. Liu K, Peng Q, Sun H, Fei M, Ma H, Hu T (In Press) A Transferred recurrent neural network for battery calendar health prognostics of energy-transportation systems. IEEE Trans Ind Inform https://doi.org/10.1109/TII.2022.3145573

34. Severson KA, Attia PM, Jin N, Perkins N, Jiang B, Yang Z, Chen MH, Aykol M, Herring PK, Fraggedakis D (2019) Data-driven prediction of battery cycle life before capacity degradation. Nat Energy 4(5):383–391

35. Liu K, Tang X, Teodorescu R, Gao F, Meng J (in Press) Future ageing trajectory prediction for lithium-ion battery considering the knee point effect. IEEE Trans Energy Convers. https://doi.org/10.1109/TEC.2021.3130600

36. Liu K, Hu X, Wei Z, Li Y, Jiang Y (2019) Modified Gaussian process regression models for cyclic capacity prediction of lithium-ion batteries. IEEE Trans Transp Electrification 5(4):1225–1236

37. Liu K, Ashwin T, Hu X, Lucu M, Widanage WD (2020) An evaluation study of different modelling techniques for calendar ageing prediction of lithium-ion batteries. Renew Sust Energ Rev 131:110017

38. Schmalstieg J, Käbitz S, Ecker M, Sauer DU (2013) From accelerated aging tests to a lifetime prediction model: analyzing lithium-ion batteries. In: Proceedings of 2013 World Electric Vehicle Symposium and Exhibition (EVS27), pp 1–12

39. Liu K, Shang Y, Ouyang Q, Widanage WD (2020) A data-driven approach with uncertainty quantification for predicting future capacities and remaining useful life of lithium-ion battery. IEEE Trans Ind Electron 68(4):3170–3180

40. Lei Y, Lin J, He Z, Zuo MJ (2013) A review on empirical mode decomposition in fault diagnosis of rotating machinery. Mech Syst Signal Process 35(1–2):108–126

41. Tang X, Liu K, Wang X, Gao F, Macro J, Widanage WD (2020) Model migration neural network for predicting battery aging trajectories. IEEE Trans Transp Electrification 6(2):363–374

42. Liu K, Li Y, Hu X, Lucu M, Widanage WD (2019) Gaussian process regression with automatic relevance determination kernel for calendar ageing prediction of lithium-ion batteries. IEEE Trans Ind Inform 16(6):3767–3777

43. Yang D, Zhang X, Pan R, Wang Y, Chen Z (2018) A novel Gaussian process regression model for state-of-health estimation of lithium-ion battery using charging curve. J Power Sources 384:387–395
44. Feng X, Ouyang M, Liu X, Lu L, Xia Y, He X (2018) Thermal runaway mechanism of lithium ion battery for electric vehicles: a review. Energy Stor. Mater. 10:246–267
45. Feng X, Pan Y, He X, Wang L, Ouyang M (2018) Detecting the internal short circuit in large-format lithium-ion battery using model-based fault-diagnosis algorithm. J Energy Storage. 18:26–39
46. Zhang M, Du J, Liu L, Siegel J, Lu L, He X, Ouyang M (2018) Internal short circuit detection method for battery pack based on circuit topology. Sci China Technol Sci 61(10):1502–1511
47. Hu X, Zhang K, Liu K, Lin X, Dey S, Onori S (2020) Advanced fault diagnosis for lithium-ion battery systems: a review of fault mechanisms, fault features, and diagnosis procedures. IEEE Ind Electron Mag 14(3):65–91
48. Lai X, Yi W, Kong X, Han X, Zhou L, Sun T, Zheng Y (2020) Online detection of early stage internal short circuits in series-connected lithium-ion battery packs based on state-of-charge correlation. J Energy Storage. 30:101514
49. Lai X, Wang S, He L, Zhou L, Zheng Y (2007) A hybrid state-of-charge estimation method based on credible increment for electric vehicle applications with large sensor and model errors. J Energy Storage 27:101106
50. Lai X, Zheng Y, Sun T (2018) A comparative study of different equivalent circuit models for estimating state-of-charge of lithium-ion batteries. Electrochim Acta 259:566–577
51. Liu L, Feng X, Zhang L, Lu X, Han X, He M uyang, Comparative study on substitute triggering approaches for internal short circuit in lithium-ion batteries. Appl Energy 259:114143
52. Liu K, Zou C, Li K, Wik T (2018) Charging pattern optimization for lithium-ion batteries with an electrothermal-aging model. IEEE Trans Ind Inform 14(12):5463–5474
53. Keil P, Jossen A (2016) Charging protocols for lithium-ion batteries and their impact on cycle life—An experimental study with different 18650 high-power cells. J Energy Storage 6:125–141
54. Ma H, You P, Liu K, Yang Z, Fei M (2017) Optimal battery charging strategy based on complex system optimization. In: Advanced computational methods in energy, power, electric vehicles, and their integration. Springer, pp 371–378
55. Liu KL, Li K, Yang ZL, Zhang C, Deng J (2016) Battery optimal charging strategy based on a coupled the thermoelectric model. In: Proceedings of IEEE Congress on Evolutionary Computation (CEC) held as part of IEEE World Congress on Computational Intelligence (IEEE WCCI), Vancouver, CANADA, pp 5084–5091
56. Wu J, Wei Z, Liu K, Quan Z, Li Y (2020) Battery-involved energy management for hybrid electric bus based on expert-assistance deep deterministic policy gradient algorithm. IEEE Trans Veh Technol 69(11):12786–12796
57. Ouyang Q, Xu G, Liu K, Wang Z (2019) Wireless battery charging control for electric vehicles: a user-involved approach. IET Power Electron 12(10):2688–2696
58. Ling Z, Zhang Z, Shi G, Fang X, Wang L, Gao X, Fang Y, Xu T, Wang S, Liu X (2014) Review on thermal management systems using phase change materials for electronic components, Li-ion batteries and photovoltaic modules. Renew Sust Energ Rev 31:427–438
59. Liu K, Li K, Zhang C (2017) Constrained generalized predictive control of battery charging process based on a coupled thermoelectric model. J Power Sources 347:145–158
60. Chen D, Jiang J, Kim G-H, Yang C, Pesaran A (2016) Comparison of different cooling methods for lithium ion battery cells. Appl Therm Eng 94:846–854
61. Qiu C, He G, Shi W, Zou M, Liu C (2019) The polarization characteristics of lithium-ion batteries under cyclic charge and discharge. J Solid State Electrochem 23(6):1887–1902
62. Gao Y, Jiang J, Zhang C, Zhang W, Ma Z, Jiang Y (2017) Lithium-ion battery aging mechanisms and life model under different charging stresses. J Power Sources 356:103–114
63. Liu K, Li K, Ma H, Zhang J, Peng Q (2018) Multi-objective optimization of charging patterns for lithium-ion battery management. Energy Convers Manage 159:151–162

64. Liu K, Li K, Yang Z, Zhang C, Deng J (2017) An advanced Lithium-ion battery optimal charging strategy based on a coupled thermoelectric model. Electrochim Acta 225:330–344
65. Zhang C, Jiang J, Gao Y, Zhang W, Liu Q, Hu X (2017) Charging optimization in lithium-ion batteries based on temperature rise and charge time. Appl Energy 194:569–577
66. Liu K, Hu X, Yang Z, Xie Y, Feng S (2019) Lithium-ion battery charging management considering economic costs of electrical energy loss and battery degradation. Energy Convers Manage 195:167–179
67. Lin X, Perez HE, Mohan S, Siegel JB, Stefanopoulou AG, Ding Y, Castanier MP (2014) A lumped-parameter electro-thermal model for cylindrical batteries. J Power Sources 257:1–11
68. Diao Q, Sun W, Yuan X, Li L, Zheng Z (2016) Life-cycle private-cost-based competitiveness analysis of electric vehicles in China considering the intangible cost of traffic policies. Appl Energy 178:567–578
69. Ouyang Q, Wang Z, Liu K, Xu G, Li Y (2019) Optimal charging control for lithium-ion battery packs: a distributed average tracking approach. IEEE Trans Industr Inform 16(5):3430–3438
70. Yan C, Fang HZ, Chao HY (2018) Battery-aware time/range-extended leader-follower tracking for a multi-agent system. In: Proceedings of American Control Conference (ACC), Milwaukee, USA, WI, pp 3887–3893

Chapter 6
Data Science-Based Battery Reutilization Management

This chapter focuses on the data science-based technologies for battery reutilization management, which is the third stage of battery full-lifespan and crucial for the sustainable development of batteries. Battery reutilization mainly includes battery echelon utilization (secondary utilization) and material recycling. During the long-term service of the battery in EVs, the consistency of the battery cell is enlarged and its health would deteriorate. Therefore, the batteries retired from EVs cannot be directly used for secondary utilization. It is necessary to evaluate the residual value of these batteries by using the historical data or the test data, and then sort and regroup them for safe secondary utilization.

In this chapter, the basic process of battery echelon utilization and material recycling is first introduced. Then, the sorting and regrouping methods based on historical or/and test data for echelon utilization are proposed and verified. Finally, some classical battery material recycling methods are discussed. Echelon utilization and material recycling are important links in the full-lifetime cycle of Li-ion battery, while building a data-based traceability platform becomes essential for efficient battery reutilization management.

6.1 Overview of Battery Echelon Utilization and Material Recycling

According to the EV world sales database, 2.65 million EVs were sold worldwide in the first half of 2021, an increase of 168% compared with 2020 (as shown in Fig. 6.1), indicating vehicle electrification is fast-growing in the world. The International Energy Agency released a report named "Global Electric Vehicle Outlook" in 2020 and predicts that the global EV ownership will continue to grow in the next 10 years. Under current policies, the number of global EVs will reach 145 million in 2030. Due to the rapid growth of EVs, the global Li-ion battery shipments reached

© The Author(s) 2022
K. Liu et al., *Data Science-Based Full-Lifespan Management of Lithium-Ion Battery*, Green Energy and Technology,
https://doi.org/10.1007/978-3-031-01340-9_6

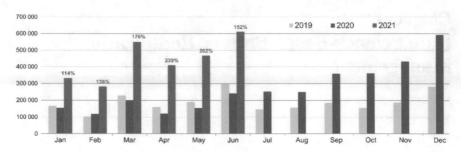

Fig. 6.1 Global sales of electric vehicles in the past 3 years

294.5GWh in 2020, and the global demand for Li-ion batteries will increase to 1.3 TWh in 2030 [1]. According to the estimation of global EVs and power batteries in 2030, it is estimated that the total retired Li-ion batteries from global EVs will reach 460 GWh in 2030, and the total mass of retired batteries will reach 12.85 million tonnes from 2021 to 2030, equivalent to 1285 Eiffel towers [2].

The massive increase in demand for Li-ion batteries has brought two serious problems [3–7]: (1) the shortage of supply chain for the battery raw material. There is a shortage of some rare metals in the upstream materials of Li-ion batteries after the large-scale mining, such as lithium, cobalt, nickel, and manganese. From 2021 to 2030, the cumulative use of cobalt in batteries will exceed 30% of the world's recoverable amount. Moreover, these rare metals are unevenly distributed in the world. For example, more than 66% of cobalt is distributed in the Democratic Republic of the Congo. Therefore, cross-regional transportation also brings risks and challenges to the supply chain. (2) The retired Li-ion batteries contain electrolytes and heavy metals. If they cannot be properly handled, great pollution will be caused to the environment, soil, and drinking water. For example, a cell with a mass of 0.2 kg can pollute 1 km^2 of land for about 50 years and enters the bodies of people and animals through the food chains [8, 9]. Therefore, the safe disposal of these retired Li-ion batteries has become a global problem that needs to be solved urgently.

In fact, the batteries retired from EVs have high economic and environmental values. On the one hand, these batteries are rich in precious metals (such as lithium, cobalt, nickel, and manganese) [7, 10], and their content in the batteries is much higher than their raw ore. If these materials are recycled from the retired Li-ion batteries, not only can the environmental pollution caused by the disposal of batteries and land burial be reduced, but more importantly, the recycled materials can be used to remanufacture batteries and reduce the mining of raw ores. Therefore, the retired Li-ion batteries can be called urban ore, which has great economic and environmental values. On the other hand, the battery retired from the EVs still has around 80% of the initial capacity. Although it does not meet the safety requirements of EVs, it can be used in other scenarios where the safety requirements are lower than that of EVs, such as energy storage stations, power supply for 5G base stations, low-speed vehicles. This method of changing application scenarios to extend the service life of the battery is called echelon utilization. It can be seen that the echelon utilization and

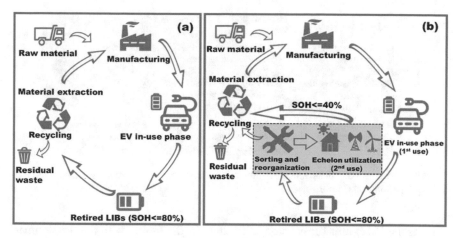

Fig. 6.2 Two technical roadmaps for the disposal of retired Li-ion batteries. **a** Material recycling; **b** Echelon utilization, reprinted from [11], with permission from Elsevier

material recycling can maximize the full lifecycle value of Li-ion batteries, reduce the application cost of Li-ion batteries, and promote the sustainable development of Li-ion batteries. The above two technical roadmaps for the disposal of retired Li-ion batteries can be described in Fig. 6.2.

6.1.1 Echelon Utilization

The best way to dispose of the retired Li-ion batteries is to perform echelon utilization first and then recycle materials, which can maximize the value of Li-ion batteries and promote the healthy and sustainable development of Li-ion batteries. Figure 6.2b shows the technical route of the echelon utilization route, which conforms to the 4R principle in concept very well: reuse, resell, update, and recycle. Therefore, this route has attracted enough attention and has also been extensively studied in recent years. With the advent of the large-scale retired Li-ion batteries and the improvement of relevant laws and regulations in various countries, the industrial chain of echelon utilization is gradually being established, and its prospects are broad.

Generally, batteries in EVs have three levels in structure: cells, modules, and packs. Therefore, the first outstanding issue in the echelon utilization is which level should be chosen for the secondary utilization, as it would directly affect the technical difficulty and costs. Figure 6.3 illustrates the advantages and disadvantages of echelon utilization at different levels, which can be summarized as:

(1) If the battery pack is disassembled to the cell level, the workload is large, the cost is high, and the battery may be damaged. Moreover, cell regrouping will generate new material costs and may bring new safety risks. Therefore, from the perspective of economy and safety, it is not recommended to use cell

Cell level	Module level	Pack level
✓ Excellent flexibility and expansibility for reuse. ✓ Good consistency after regrouping. ✓ High cell utilization.	✓ Easy to disassemble, no damage to battery. ✓ Historical data available in BMS. ✓ Good flexibility and expansibility.	✓ Low cost and high economy.
✗ High disassembly cost. ✗ The disassembled cell is easy to break.	✗ Some modules can not be reused due to the defects of some cells.	✗ Only suitable for specific scenes. ✗ Poor consistency.

Fig. 6.3 Advantages and disadvantages of the echelon utilization at different levels, reprinted from [11], with permission from Elsevier

as the basic unit for echelon utilization. In the future, with the development and maturity of robot automatic disassembly technologies, cells become more possible to be regrouped, which is worthy of recommendation due to its high flexibility.

(2)   Echelon utilization at the pack level is the most economical solution because it does not require disassembly. However, some of the cells in the retired battery pack have poor consistency, and the safety of secondary use is poor. Finally, the utilization at the pack level is limited by the application scenarios.

(3)   The echelon utilization at the module level is the best choice at present, due to the advantages of economy and flexibility. It can screen out the poor-quality modules without the requirements of deep disassembly for improving the safety of the regrouping system, further presenting good regrouping flexibility.

Then, the second outstanding problem is how to disassemble battery pack to the module. Battery disassembly is a complicated and dangerous task. First, the battery is a high-energy and chemical carrier; if it is not handled properly, it will cause various safety issues such as short circuits and liquid leakage. What's more serious is that it may cause an explosion or fire, further leading to casualties and property damage. Second, the toxic and carcinogenic substances in battery electrolyte and electrode materials may cause chemical hazards during disassembly. Third, batteries from different manufacturers may have different external structures, module connection methods, and process technologies, further bringing difficulties to the automatic disassembly. Therefore, battery disassembly is mostly manual or semi-automatic at present. The establishment of flexible and intelligent battery disassembly equipment is an urgent need for development in the future. In summary, the disassembly of large-scale retired Li-ion batteries faces the following technical challenges: (1) Lack of skilled dismantling workers and professional demolition tools to ensure the safety, integrity, and speed of disassembly; (2) The disassembly efficiency is low, further directly affecting the economic value; (3) The high reliability of battery component connection increases the difficulty and time cost of battery disassembly. A feasible solution is to standardize the design of cells, modules, and packs, and to design a battery connection method that is conducive to disassembly. In addition, the development of automatic disassembly robots is essential. However, due to the complexity of battery structure and connectivity, and rich wires and battery management system, complete automatic disassembly becomes a great challenge, and man–machine integration is a short-term solution.

The third outstanding issue is how to evaluate the residual value of the retired Li-ion batteries, and how to sort and regroup them for secondary use. The most critical issue for sorting the retired Li-ion batteries is how to ensure its accuracy and efficiency, which would directly determine the economy and safety of the echelon utilization. Regrouping should not only be achieved based on the sorting results, but also based on actual application scenarios. For example, high-capacity retired Li-ion batteries are suitable for energy applications, while low internal resistance Li-ion batteries are more suitable for power applications. For the retired Li-ion batteries without historical data, it is very time-consuming and energy-consuming to test batteries one by one. Obviously, the traditional sorting method based on capacity or internal resistance test is not applicable. An important challenge for the echelon utilization of the large-scale retired Li-ion batteries is to design the sorting methods and devices with excellent sorting speed and accuracy, which is suitable for different application scenarios. For the retired Li-ion batteries with historical data, the technical challenge here becomes how to quickly sort through massive amounts of data. At present, most retired Li-ion batteries have no historical data or the quality of historical data is poor. In this case, how to sort Li-ion batteries is challenging, while some data science-based methods will be described in detail in Sect. 6.2.

6.1.2 Material Recycling

With the innovation of battery material systems and the improvement of environmental protection requirements, battery recycling technology is facing major opportunities and severe challenges. First, since the cathode material is the most valuable one in a battery, the "dissolution–precipitation–recycling" process of valuable metals (such as cobalt) has formed a certain economic recovery-driven business model. However, to reduce battery costs and increase energy density, cathode materials are constantly being updated, and the research on waste Li-ion battery recycling technology is relatively backward, making it difficult for the corresponding recycling technology to keep up with the pace of material updates. If recyclers cannot recycle pure and high-quality materials, the recyclable value will become very low. Moreover, ternary batteries are developing in the direction of high nickel and low cobalt, which will weaken the existing battery recycling business model. Therefore, new recycling technologies need to be continuously updated and iterated to meet market demand.

Second, most of the current recycling technologies focus on the recycling of valuable metals such as lithium, nickel, cobalt, and copper, and little attention is paid to other low-value components. However, the recycling of negative electrode materials and electrolytes is an unavoidable issue. Here the electrolyte is volatile and generates gases such as HF and PF5, which are harmful to both the human body and the environment. In recent years, the recycling and reuse of electrolyte have become a research hotspot. However, these methods still have problems such as long recycling process, high recycling cost, low recycling rate, making them become unsuitable for large-scale industrial production, and lack market driving force. Third, the current recycling process of Li-ion batteries usually involves battery disassembly, smelting and/or acid leaching, chemical precipitation separation, and decomposition. In these processes, a large amount of energy and chemicals will be consumed, causing greenhouse gas emissions, secondary waste, and other environmental problems. Therefore, the development of green, efficient, and full-component recycling methods becomes a future trend.

The material recycling and resource utilization of waste Li-ion batteries can help recover the valuable materials, realize the recycling of valuable resources, reduce the impact of waste treatment on the environment, and reduce the development and consumption of natural resources. Li-ion batteries usually consist of a cathode, an anode, an electrolyte, a separator, a casing, and other components. A typical Li-ion battery contains about 25–30% cathode, 15–30% anode, 10–15% electrolyte, 18–20% shell, 3–4% independent components, and 10% other components. Typical cathode material is composed of 80–85% metal oxide powder, about 10% polyvinylidene fluoride binder, and 5% acetylene black. Graphite is commonly used as a negative electrode material, and it consists of hexagonal carbon atoms arranged in thin sheets. The separator is usually made of microporous polypropylene or polyethylene. The commonly used electrolytes in Li-ion battery include $LiClO_4$, $LiAsF_6$, $LiPF_6$, $Li(CF_3SO_3)$, $Li[N(CF_3SO_2)_2]$. It can be seen that the cathode material is the most valuable part. Table 6.1 lists the chemical components of several typical Li-ion batteries. Obviously, the metal content of Li-ion batteries is even better than that of natural raw materials. According to the statistics from London Metal Exchange (LME), the average price of major metal materials in Li-ion batteries in 2020 is shown in Table 6.2, indicating that the metal materials in Li-ion batteries have a high recycling value, of which cobalt has the highest value. Therefore, high-cobalt Li-ion batteries have excellent recycling value. Ternary batteries have a higher recycling value than lithium iron phosphate (LFP) batteries.

There are many topics about battery material recycling, such as recycling mode and industrial chain, recycling methods. In this chapter, it should be noted that in the study of echelon utilization, the battery is called retired battery, while in the discussion of battery recycling, it is called waste or spent battery.

Table 6.1 Chemical composition of some typical Li-ion batteries

Metals	Cathode material				
	$LiCoO_2$ (mass%)	$LiFePO_4$ (mass%)	$LiMn_2O_4$ (mass%)	$LiNi_{1/3}Mn_{1/3}Co_{1/3}O_2$ (mass%)	$LiNi_{0.8}Co_{0.15}Al_{0.05}O_2$ (mass%)
Aluminium	5.2	6.5	21.7	22.72	21.9
Cobalt	17.3	0.0	0.0	8.45	2.3
Copper	7.3	8.2	13.5	16.60	13.3
Iron/Steel	16.5	43.2	0.1	8.79	0.1
Lithium	2.0	1.2	1.4	1.28	1.9
Manganese	0.0	0.0	10.7	5.86	0.0
Nickel	1.2	0.0	0.0	14.84	12.1
Binder	2.4	0.9	3.7	1.39	3.8
Electrolyte	14.0	14.9	11.8	11.66	11.7
Plastic	4.8	4.4	4.5	3.29	4.2

Table 6.2 Price of main metal materials in Li-ion batteries (LMC, 2020)

Metals	Prices ($/kg)	Metals	Prices ($/kg)
Cobalt	49.81	Aluminium	2.36
Nickel	16.16	Iron	0.73
Copper	9.456	Phosphate rock	0.074
Manganese	2.30	Lithium hydroxide	11.75

6.2   Sorting of Retired Li-Ion Batteries Based on Neural Network

The sorting and regrouping of the retired Li-ion batteries are based on one or more battery performance criteria, named sorting criteria. The purpose of battery sorting is to evaluate the residual value of the battery through these criteria. As most retired Li-ion batteries do not have full historical data, the evaluation of battery residual value can only be done through testing. In this section, the following two outstanding issues for data science-based battery sorting are addressed: (1) What kind of criteria should be constructed to truly reflect the ageing characteristics of the retired battery; (2) How to quickly and accurately obtain these criteria based on the test data for the sorting of the large-scale retired Li-ion batteries.

6.2.1   Data Science-Based Sorting Criteria

The consistency of Li-ion batteries retired from EVs is usually poor. Therefore, it is necessary to sort the retired Li-ion batteries before echelon utilization. In other words, Li-ion batteries with the same or similar performance need to be classified into the same category. The retired Li-ion battery may be in an accelerated period of performance degradation, and the inconsistency of cells or modules may increase more dramatically, further seriously affecting battery safety. Therefore, improving the accuracy and efficiency of sorting is of great significance to enhance the safety and economy of echelon utilization. The first basic but key problem of the data science-based sorting is to select one or more sorting criteria that can accurately reflect the true state of the retired Li-ion batteries.

There are many performance parameters that can characterize the residual value and health status of retired Li-ion batteries. Figure 6.4 lists the common sorting criteria at different scales. Theoretically, the performance degradation of retired Li-ion batteries is caused by the microscopic changes in the structure or material morphology at the molecular scale. Therefore, the microscale sorting criteria become most direct and accurate. However, it has a cumbersome process and requires high-precision equipment. Thus, it becomes unsuitable for the sorting of large-scale retired Li-ion batteries. The sorting criteria widely used in literature include battery appearance [12, 13], capacity or life [14–17], internal resistance [17], impedance spectrum

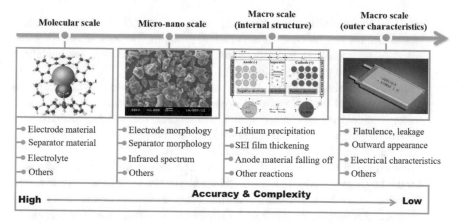

Fig. 6.4 Sorting criteria for the retired Li-ion batteries at different scales, reprinted from [11], with permission from Elsevier

[18, 19], charge and discharge curve characteristics [12, 13], or their combination. However, these macro characteristics can only reflect the battery external macro-characteristics and cannot reflect the internal state of the battery. Therefore, there is a contradiction between precision and complexity in the selection of sorting indicators, and how to balance them is a key issue.

To build suitable criteria for the large-scale retired Li-ion batteries, the following issues need to be considered:

(1) The multidimensional sorting criteria need to be constructed to evaluate battery status more comprehensively. The current single sorting index cannot fully reflect the ageing characteristics of Li-ion batteries. For example, classifying the retired Li-ion batteries just based on capacity may result in the batteries with different internal resistances being classified into the same category, which is obviously unreasonable. More importantly, the capacity and internal resistance cannot fully characterize the safety characteristics of a battery, and safety becomes most important in secondary utilization. In the sorting process, the typical side reactions (such as lithium plating, thickening of SEI film) or typical failures (such as internal short circuits) that seriously threaten the battery safety should be considered. Moreover, the secondary battery is in a period of accelerated decline in life, the inconsistency between battery modules or cells may be enlarged gradually, and capacity diving may occur, which will threaten the battery safety. Therefore, the prediction of battery life trajectory becomes significantly important. In summary, the ideal battery residual value evaluation requires the construction of multidimensional indicators, such as the typical side effects (representing the past states of Li-ion batteries), capacity and internal resistance (representing the current states), and life trajectory (representing the future states). These indicators can comprehensively evaluate the status of Li-ion batteries from different dimensions and improve the

safety of echelon utilization. However, how to quickly obtain the multidimensional criteria has become an outstanding problem. The modern sensing and detection technologies, network technologies, big data science, and modelling technologies are powerful solutions.

(2) The available historical data of the battery is very valuable, which contains an abundance of battery status information. Using historical data to evaluate battery residual value is an economical and accurate method. However, most of the existing historical data is unlabelled, further bringing challenges to battery state estimation and is also an issue to be solved.

(3) The sorting efficiency is directly related to the economy and must be considered for the large-scale retired Li-ion batteries. Currently, most of the retired Li-ion batteries do not have available historical data, the sorting criteria can only be obtained through the test data of Li-ion batteries in this case. The traditional measurement method of the battery capacity and internal resistance requires about 3 h, which is not allowed for the state estimation of the large-scale retired Li-ion batteries. How to quickly obtain the criterion based on the test data becomes important. In Sect. 6.2.2, a valuable data science-based technical route of the fast capacity estimation is introduced.

However, the capacity and internal resistance are the external performance of the battery and cannot well characterize battery internal states. Electrochemical impedance spectroscopy (EIS) is a non-destructive measurement method that has been widely used to evaluate the performance of Li-ion batteries. EIS can reflect the internal characteristics of a battery with the minute level of test time. Therefore, it is a promising method for evaluating the residual value of retired Li-ion batteries. However, EIS is susceptible to factors such as SoC, SoH, charge/discharge rate, and temperature. In this context, further exploration is needed in the sorting process and evaluation. An effective data science-based method is proposed in Sect. 6.2.3.

6.2.2 Case 1: Sorting Criteria Estimation Based on Charging Data

To evaluate SoH of reutilized Li-ion batteries, capacity and internal resistance testing is the most direct method. However, this testing is very time-consuming and energy-consuming. In this section, a fast data science method of estimating reutilized battery capacity and internal resistance based on charging curves and machine learning is introduced.

Figure 6.5 illustrates the principle of the proposed data science method. It can be expressed as follows: First, a large quantity of retired Li-ion batteries (battery modules or cells) to be sorted are connected in parallel and rest for a long time to ensure that the battery terminal voltage becomes consistent. This process can be carried out in large batches and is obviously efficient. Then, these batteries are randomly selected for series charging, and a series of short-term charging voltage

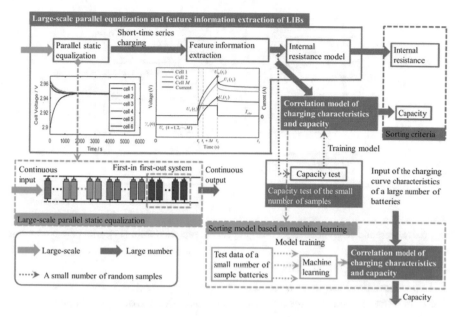

Fig. 6.5 A fast estimation method of capacity and internal resistance based on machine learning, reprinted from [11], with permission from Elsevier

curves are obtained with rapid speed. These curves can be used to extract internal resistance and capacity. Afterwards, the capacity of a small number of samples is obtained through the standard capacity test, and a machine learning algorithm (such as neural network) will be used to establish a correlation model among the charge–discharge curve, internal resistance, and capacity. Finally, the established correlation model can be used to estimate the internal resistance and capacity characteristics of the remaining massive batteries. In this method, states of massive batteries will be estimated from a small sample test using machine learning, which greatly improves the sorting efficiency. However, how to establish a black-box model for accurately correlating charging/discharging curves with battery capacity has become a key issue.

Take the sorting of battery cells as an example to illustrate the process of obtaining capacity and internal resistance. The process of obtaining capacity and internal resistance is as follows: First, a large number (N) of cells are selected for the parallel equalization to ensure that the terminal voltage of the cells remains the same. Then, M cells ($M \leq N$) are selected for the series constant-current charging. Figure 6.6 shows the voltage curves of M cells in the serial-charging process. Simply, the internal resistance of each cell can be calculated as follows:

$$R_{\text{cha}, \Delta t} = \frac{U_k(t_1 + \Delta t) - U_k(t_1)}{I_{\text{cha}}} \tag{6.1}$$

where $R_{\text{cha}, \Delta t}$ is the charging internal resistance at time Δt; I_{cha} is charging current.

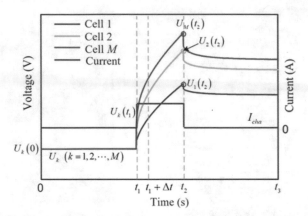

Fig. 6.6 Schematic diagram of series charging curves, reprinted from [17], with permission from Elsevier

In addition, the charging curves of each cell will be affected by internal resistance. Therefore, the residual voltage after excluding the influence of internal resistance can be expressed as:

$$\tilde{U}_k = U_k(t_2) - (U_k(t_1 + \Delta t) - U_k(t_1)) \tag{6.2}$$

where $\tilde{U}_k$ is the terminal voltage of cell k ($k = 1, 2, \ldots, M$) excluding the influence of internal resistance; $U_k(t_1), U_k(t_1 + \Delta t)$, and $U_k(t_2)$ are the terminal voltage at $t_1, t_1 + \Delta t$, and t_2, respectively.

Then, a small number of cells from M cells are randomly selected for the standard capacity test, and then the test capacities (C_k) and $\tilde{U}_k$ are used as samples to train the capacity model based on machine learning. In this study, a back-propagation neural network (BPNN) is applied to get the capacity model. Since the BPNN belongs to a frequently reported algorithm, it will not be described in detail here for brevity.

To verify the effectiveness of the proposed method, 108 retired cells are used to carry out a fast capacity estimation. After retiring a battery from EVs, its SoC would be generally discharged below 30% to ensure its safe transportation. Therefore, the SoCs of the experimental cells would be adjusted to different SoC levels for constructing the following three cases: Case 1: SoC = 5%, Case 2: SoC = 20%, Case 3: SoC = 30%. Then, the 108 cells are connected in series for the constant-current charging to get a series of charging curves. In the above three cases, the methods described in Sect. 6.2.1 are used to estimate battery capacity with the results shown in Fig. 6.7. Note that the capacity model is trained based on 36 battery cells, while another 72 cells are used to verify the accuracy of capacity estimation.

It can be observed that the capacity estimation errors in these three cases are all within 5%. It can be inferred that the accuracy of capacity estimation will be further improved as the number of sample batteries increases. In the proposed data science method, a machine learning algorithm is used to establish a black-box model

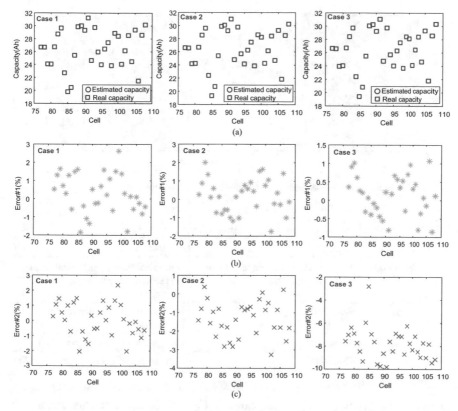

Fig. 6.7 Capacity estimation results in different cases. **a** Capacity estimation results; **b** capacity estimation error, reprinted from [17], with permission from Elsevier

between battery capacities and charging curves, which greatly improves the efficiency of capacity estimation. It is very valuable for the sorting of large-scale retired Li-ion batteries.

6.2.3   Case 2: Sorting Criteria Estimation Based on EIS

6.2.3.1   Methodology

To accurately estimate the ageing characteristics of the retired battery, EIS is used for the non-destructive testing of Li-ion batteries. Here EIS test belongs to an efficient in-situ/ex-situ electrochemical characterization technology [20], which has been widely used in the electrochemical measurement and characterization of battery ageing [21, 22]. The EIS information can be expressed in the form of a Nyquist plot combining the real and imaginary parts of the impedance. The Nyquist plot of the impedance

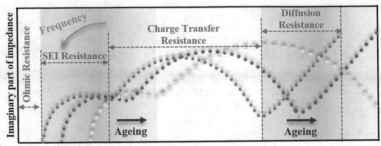

Fig. 6.8 Frequency distribution at Nyquist impedance spectroscopy for a cell

Table 6.3 Physical meaning of parameters in EIS diagram

Parts	Frequency	Features	Impedance
I	Ultrahigh	The intersection of impedance and the horizontal axis	Ohmic resistance (R_0): Resistance of the cell (electrolyte, separator, and electrodes)
II	High	Semicircle	Resistance of solid electrolyte interphase (R_{SEI})
III	Middle	Semicircle	Charge transfer impedance (R_{ct})
IV	Low	45° straight line	Lithium-ion diffusion impedance (R_d)

spectrum of a Li-ion battery cell is shown in Fig. 6.8, in which the horizontal axis is the real part of the impedance, and the vertical axis is the imaginary part of the impedance. The physical meanings of each parameter in Fig. 6.8 are listed in Table 6.3 [23, 24].

Reference [25] pointed out that if the high-frequency part of the impedance spectrum, that is, the second semicircle part, is directly fitted, the data distortion may be caused, and the distribution of relaxation times (DRT) method can effectively avoid this problem. The AC impedance of Li-ion batteries is written as follows:

$$z(f) = R_0 + \int_0^\infty \frac{\gamma(\tau)}{1 + j2\pi f \tau} d\tau + \frac{1}{j2\pi f c_{in}} \tag{6.3}$$

where f is the frequency, γ is the time distribution of polarization losses, τ is the time constant of the corresponding impedance, and c_{in} is the intercalation capacitance. The main purpose of DRT is to determine $\gamma(\tau)$ through $z(f)$.

Figure 6.9a shows the EIS test results of 35 retired cells with 20% SoC at 25 °C. Moreover, as shown in Fig. 6.9b, the DRT curve could be divided into four intervals [25], which is denoted as S_1, S_2, S_3, and S_4. Here the DRT curve can reflect different impedance types, as listed in Table 6.4. The magnitude of different impedances can be calculated by the upper and lower limits of the time constants of different

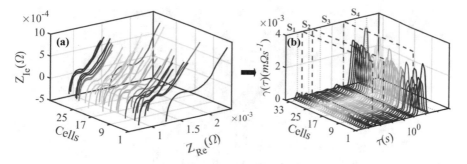

Fig. 6.9 EIS test results of 35 retired cells with 20% SoC at 25 °C. **a** Nyquist plots; **b** DRT plots

Table 6.4 Types of impedance reflected in the different DRT intervals of Li-ion batteries

Intervals	Impedance
S_1	Contact resistance (R_c)
S_2	SEI resistance (R_{SEI})
S_3	Charge transfer resistance (R_{ct})
S_4	Diffusion resistance (R_d)

impedances. For each interval, the polarization resistance R_p could be calculated as follows:

$$R_p = \int_{\tau_L}^{\tau_U} \gamma(\tau)\,d\tau \qquad (6.4)$$

where τ_U and τ_L are the upper and lower limits of time constant, respectively.

Battery capacity is one of the key criteria for the sorting of retired Li-ion batteries. However, traditional battery capacity is generally obtained by the standard capacity test, which requires a battery to be fully charged and discharged three times. Therefore, this test process is very time-consuming with the hour level, becoming obviously not conducive to the sorting of large-scale retired Li-ion batteries. The EIS test is very fast at the minute level, and the test results contain rich electrochemical characteristics. However, there is no direct relationship between capacity and EIS. In this study, an EIS-capacity correlation model is built using BPNN. As shown in Fig. 6.10, the brief process can be summarized as: First, a small number of sample cells are selected for the standard capacity test. Then, the capacity obtained from the capacity test and the DRT characteristics are used to train the BPNN model, and the EIS-capacity correlation model is obtained. Finally, the EIS-capacity correlation model is used to estimate the capacity of a large number of remaining cells. In this process, the capacity of cells can be obtained quickly only by inputting DRT characteristics. It can be seen that the advantage of the proposed method is that the capacity of most cells can be estimated quickly and accurately based on the capacity test of a

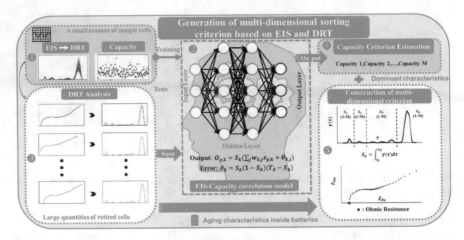

Fig. 6.10 Establishment process of the correlation model between capacity and EIS

small number of cells. In addition, other valuable battery ageing characteristics can be quickly extracted from EIS, such as S_1, S_2, S_3, and S_4. These electrochemical characteristics can reflect the internal state of the battery, which becomes useful for evaluating the residual value of the retired Li-ion batteries.

6.2.3.2 Experimental Verification

In this study, the EIS tests are conducted on 35 retired cells. Then, the capacity of these cells is tested for the reference of the fast capacity estimation, and the results are shown in Fig. 6.11. In general, the retired Li-ion batteries are transported and stored at low SoC for safety, and the previous studies [26, 27] showed that the Nyquist plots of the same Li-ion batteries with different ageing degrees at low SoC are more distinct than those at high SoC. Therefore, the SoC of each retired cell is adjusted to 20% for the EIS tests.

In this study, the DRT (DRT_i) and capacity $C_i^* (i = 1, 2, \ldots, 30)$ of 30 cells are randomly selected as the training set of the BPNN model, and the rest five cells are selected as the validation set. Owing to the DRT of each cell having the same time constant, DRT_i can be expressed as follows:

Fig. 6.11 Capacity distribution of the 35 retired cells

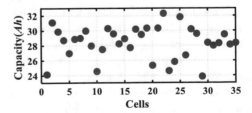

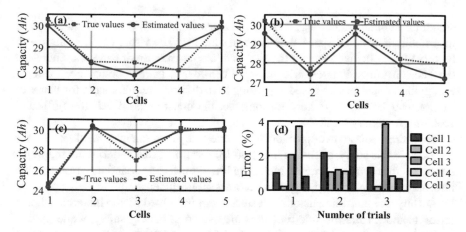

Fig. 6.12 Capacity estimation results based on the EIS-capacity correlation model

$$\widetilde{DRT}_i = \gamma_i(\tau) \quad (i = 1, 2, \ldots 30) \tag{6.5}$$

Owing to the randomness of the initial value, the training results of the BPNN model are generally unequal. In this study, the EIS-capacity model is trained three times, and the results of the verification set are shown in Fig. 6.12. It can be observed that the capacity estimation errors are less than 4% under the three training models. Therefore, it can be concluded that the EIS-capacity correlation model is accurately established using BPNN. Moreover, the EIS test time of each cell can be controlled within 10 min, while the traditional capacity test requires 3 h. It can be clearly seen that the efficiency of capacity acquisition of the proposed method is ten times higher than that of the traditional capacity test method with satisfactory accuracy, which greatly improves the sorting efficiency of the large-scale retired Li-ion batteries.

6.3 Regrouping Methods of Retired Li-Ion Batteries

6.3.1 Overview of Regrouping Methods

After obtaining the performance evaluation criteria of the retired Li-ion batteries, the next key question is how to regroup these Li-ion batteries based on the sorting criteria. The following issues need to be considered: (1) The performance evaluation of the battery involves multiple criteria, such as capacity, internal resistance, and EIS characteristics. Mathematically, the battery regrouping is a clustering issue. There are many current data science-based clustering algorithms, such as K-means clustering [28, 29], mean-shift clustering [30], density-based spatial clustering with noise [31], expectation–maximization clustering [32], cohesive hierarchical clustering [33], support vector machine regression analysis [34]. However, most of them are suited to binary classification for two-dimensional data. How to cluster batteries in a multidimensional space thus becomes the first key issue. (2) The purpose of battery regrouping is to ensure the safe and long-term reuse of batteries. Therefore, the echelon utilization scenario must be considered in regrouping. For example, energy-based and power-based scenarios have different requirements for batteries. The second key issue is how to construct the constraints of echelon utilization scenarios in battery clustering.

To address these two problems, an effective data science-based solution is provided in this study, as shown in Fig. 6.13. Assuming that batteries are regrouped based on four sorting criteria including the typical side reactions, residual life, internal resistance, and capacity, this becomes a four-dimensional clustering problem. Frequently, the multidimensional clustering can be solved by the hierarchical clustering method. Here the side reactions are related to battery safety, while the life, internal resistance, and capacity characteristics are related to battery functionality. For the echelon utilization of retired Li-ion batteries, battery safety is the priority. Therefore, the first level of battery clustering can use the side reaction characteristics as a criterion, which is a one-dimensional clustering problem. The purpose of the

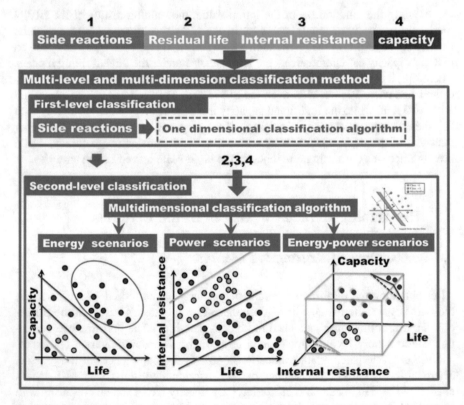

Fig. 6.13 Multidimensional clustering method under echelon utilization scenario constraints, reprinted from [11], with permission from Elsevier

clustering is to classify Li-ion batteries with the same or similar side reaction characteristics. Li-ion batteries with the same side reactions are then clustered at the second level based on the residual life, capacity, and internal resistance, making it become a three-dimensional clustering problem. Furthermore, three-dimensional clustering can be further transformed into two-dimensional clustering by using constraints of echelon utilization scenarios. Generally, echelon utilization scenarios include energy-based and power-based scenarios. The typical application of the former is energy storage power station, while that of the latter is low-speed vehicles. Obviously, energy-based scenarios require high-capacity batteries, while power-based scenarios require low-resistance batteries. In this context, batteries used in energy-based application scenarios can be clustered by capacity and residual life, while batteries used in power-based scenarios can be clustered by internal resistance and residual life. They are two-dimensional clustering problems, which are easy to implement through algorithms and easy to understand.

It should be noted that the four sorting criteria listed above are just an example. There are many criteria for clustering retired Li-ion batteries, such as capacity, internal resistance, and EIS characteristics. Two cases are given below to illustrate the application of the data science-based clustering method in the regrouping of the retired Li-ion batteries.

6.3.2 Case 1: Hard Clustering of Retired Li-Ion Batteries Using K-means

In this study, the K-means algorithm is adopted to cluster and regroup modules to improve the overall consistency of the entire battery pack. K-means is a well-known data science-based clustering algorithm in which the data items are clustered into K clusters such that each item only blogs to one cluster. Moreover, various echelon utilization scenarios have different requirements for capacity and internal resistance consistency. For example, greater capacity consistency is required than internal resistance consistency in an energy-type echelon application (e.g. energy storage power station), whereas more attention should be paid to the internal resistance consistency in a power-type application (such as the low-speed EVs). The energy-type and power-type echelon applications are two representative echelon utilization scenarios. Therefore, a reasonable data science-based clustering algorithm should consider the utilization scenario constraints.

The improved K-means algorithm is developed to describe the constraints of the echelon utilization scenarios. The proposed algorithm flow is listed in Table 6.5, in which the echelon utilization scenario factor δ is introduced, as shown in Eq. (6.9). Specifically, the closer the δ tends to 1, the higher the capacity consistency of the battery, which is accordingly clustered as being more suitable for an energy-type scenario with strict capacity requirements. The closer δ tends to 0, the higher the power density consistency of the battery, which is accordingly clustered

Table 6.5 Process of the improved K-means algorithm

I. Data standardization

The sample set is described as follows:

$$\begin{cases} A = \{x_1, x_2, \ldots x_m\} \\ x_i = [x_{c,i}, x_{r,i}] \ (i = 1, 2, \ldots, m) \end{cases} \quad (6.6)$$

where x_i is a two-dimensional array of internal resistance and capacity, and $x_{c,i}$ and $x_{r,i}$ are the capacity and internal resistance data of the ith retired cell, respectively

Then, the data is standardized as follows:

$$x_i = \left[\frac{x_{c,i} - x_c^{\min}}{x_c^{\max} - x_c^{\min}}, \frac{x_{r,i} - x_r^{\min}}{x_r^{\max} - x_r^{\min}} \right] \quad (6.7)$$

where $x_c^{\min}$, $x_c^{\max}$, $x_r^{\min}$, and $x_r^{\max}$ are the minimum capacity, maximum capacity, minimum internal resistance, and maximum internal resistance, respectively

II. Algorithm initialization

• The number of clusters is set to K, and the maximum number of iterations is set to N

• The k samples are randomly selected in sample set A as the initial clustering centre, which is described as follows:

$$\begin{cases} B = \{\mu_1, \mu_2, \cdots, \mu_k\} \\ \mu_j = [\mu_{c,j}, \mu_{r,j}] \ (j = 1, 2, \ldots, k) \end{cases} \quad (6.8)$$

where $\mu_{c,j}$ and $\mu_{r,j}$ are the capacity and internal resistance of the kth sample, respectively

III. Clustering

(1) For each element x_i in the sample set A, the Euclidean distance d_{ij} between x_i and μ_j is calculated. Here, the echelon utilization scenario factor is considered in the Euclidean distance calculation, which is described as follows:

$$d_{ij} = \sqrt{\delta^2 \left(x_{c,i} - \mu_{c,j}\right)^2 + (1 - \delta)^2 \left(x_{r,i} - \mu_{r,j}\right)^2} \quad (6.9)$$

where δ is defined as the echelon utilization scenario factor ($0 < \delta < 1$)

(2) For $j = 1, 2, \ldots, K$, the new clustering centre μ_j is updated as follows:

$$\mu_j = \frac{1}{|C_j|} \sum_{x \in C_j} x \quad (6.10)$$

(3) Repeat Steps (1) and (2) until the termination condition is met. The termination condition is defined as μ_j no longer changes or reaches the maximum number of iterations N

IV. Output clustering results

as being more suitable for a power-type scenario with high-power requirements. By setting different δ, various echelon utilization scenarios can be described, and more reasonable clustering results can be obtained.

To verify the effectiveness of the proposed regrouping method, clustering is performed using both capacity and internal resistance of 108 cells obtained in Sect. 6.2.2. To highlight the advantages of the proposed method, the conventional K-means and the improved K-means algorithms are used to cluster these cells, and the results are shown in Figs. 6.14 and 6.15, respectively. It can be observed that cells are divided into six categories, and different values of δ will lead to different clustering results. Specifically, the smaller the value of δ, the more consistent the internal resistances of the cells in the same group, and the less consistent the battery capacities. Therefore, the cluster results for a small δ are beneficial for a power-type

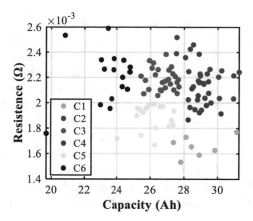

Fig. 6.14 Cluster result based on the conventional K-means algorithm, reprinted from [29], with permission from IEEE

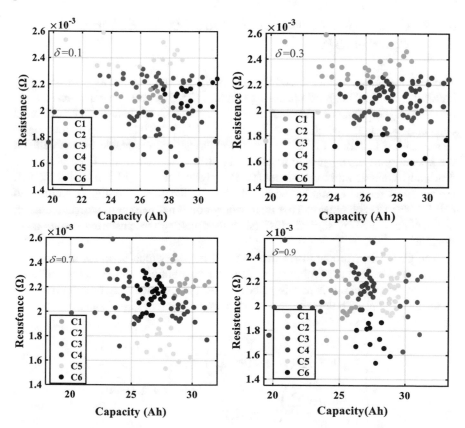

Fig. 6.15 Cluster results under different coefficient δ, reprinted from [29], with permission from IEEE

echelon utilization scenario. In contrast, the closer the value of δ to 1, the more consistent the capacities of the cells in the same cluster, and the less consistent the battery internal resistances. Therefore, the cluster results for a large δ are beneficial for an energy-type echelon utilization scenario. Simply stated, cluster results can be obtained with emphasis on either capacity or internal resistance by setting the echelon utilization scenario coefficient δ to provide cluster results that are more suitable for the intended echelon utilization scenarios.

It should be noted that the number of clusters is determined before clustering, which directly affects the clustering results. In the actual echelon utilization, the number of clusters can be determined according to the number of batteries, capacity, and internal resistance distribution. The regrouping of retired Li-ion batteries based on the cluster results can improve the adaptability and safety of cascade utilization scenarios.

6.3.3   Case 2: Soft Clustering of Retired Li-Ion Batteries Based on EIS

The traditional data science-based linear clustering methods, such as K-means, hierarchical clustering, are hard-clustering methods [35, 36], which means each sample is only assigned to a specific cluster. For the retired Li-ion batteries, the clustered cells only belong to one cluster, which is obviously inflexible for the regrouping of retired Li-ion batteries. For example, it is obviously unreasonable for the cells on the boundary of the clustering results to be strictly restricted to a fixed cluster. In this study, Gaussian mixture model (GMM) algorithm [37, 38] is used for the soft clustering of retired Li-ion batteries, which gives the possibility that one clustered cell belongs to each cluster. That is, a battery could not belong to a certain cluster, but may be shared by several clusters, further making the clustering results more flexible.

The GMM is a probabilistic data science model, which assumes that all data points are generated by a mixture of Gaussian distributions with a finite number of unknown parameters. It is usually used for unsupervised learning or soft clustering of unlabelled data. The probability distribution function of GMM can be defined as follows:

$$
\begin{cases}
\pi(x) = \sum_{k=1}^{K} p_k N\left(x \mid \mu_k, \sum_k\right) \\
N\left(x \mid \mu_k, \sum_k\right) = \dfrac{1}{(2\pi)^{\frac{d}{2}} \left|\sum_k\right|^{\frac{1}{2}}} e^{\left[-\frac{1}{2}(x-\mu_k)^T \sum_k^{-1}(x-\mu_k)\right]} \\
\sum_{k=1}^{K} \pi_k = 1
\end{cases}
\tag{6.11}
$$

where x is the data points, K is the number of clusters, N is the Gaussian density function, π is the mixing probability, μ_k and $\sum_k$ are the mean and covariance for the Gaussian k ($k \in \{1, \ldots, K\}$), respectively, and d is the data dimension.

Generally, the expectation–maximization algorithm is an iterative algorithm that is used to find the maximum likelihood estimation of the GMM when the parameters cannot be found directly, and it can be simply divided into the expectation step (E-step) and the maximization step (M-step). In the E-step, the available data is used to estimate the value of missing variable. Based on the estimated value, the parameters are updated with the complete data. Suppose the parameter to be determined is $\theta = [\pi, \mu, \Sigma]$, and the detailed process of the GMM are listed in Table 6.6.

The silhouette value $s(x_i)$ and silhouette coefficient C_{SC} are used to evaluate the clustering effects, and they can be expressed as follows [39]:

$$s(x_i) = \frac{b(x_i) - a(x_i)}{\max(a(x_i), b(x_i))} \tag{6.18}$$

$$C_{SC} = \frac{1}{m} \sum_{i=1}^{m} s(x_i) \tag{6.19}$$

Table 6.6 Process of the GMM algorithm

Name	Content
E-step	According to the current model parameters, calculate and evaluate the probability that the observation data x_i belongs to the cluster C_k as follows: $Q(\theta^*, \theta^{(t)}) = \mathbb{E}[\ln p(X, Z\|\theta^*)]$ $= \sum_{k=1}^{K} \sum_{i=1}^{N} [\ln \pi_k + \ln N(x_i\|\mu_k, \Sigma_k)] p(z_i = 1\|x_i, \theta^{(t)})$ (6.12) where θ^* is the revised parameters, and Z is the all possible latent variables
M-step	Determine the parameters by using the maximum likelihood, and find the revised parameters as follows: $\theta^* = \arg\max Q(\theta^*, \theta^{(t)})$ (6.13) To determine the optimal parameter π, take the derivative of $Q(\theta, \theta^*)$ with π and then set it equal to zero: $\frac{\partial Q(\theta,\theta^*)}{\partial \pi_i} = \sum_{i=1}^{N} \frac{p(z_i=1\|x_i,\theta^{(t)})}{\pi_i} - \lambda\left(\sum_{k=1}^{K} \pi_k - 1\right) = 0$ (6.14) where λ is the Lagrange multiplier From Eq. (14), we obtain: $\pi_k^* = \frac{\sum_{i=1}^{N} p(z_i=1\|x_i,\theta^{(t)})}{N}$ (6.15) Similarly, the optimal parameters μ_k^* and Σ_k^* are obtained: $\mu_k^* = \frac{\sum_{i=1}^{N} (p(z_i=1\|x_i,\theta^{(t)})x_i)}{\sum_{i=1}^{N} p(z_i=1\|x_i,\theta^{(t)})}$ (6.16) $\Sigma_k^* = \frac{\sum_{i=1}^{N} (p(z_i=1\|x_i,\theta^{(t)}) \cdot (x_i-\mu_k)(x_i-\mu_k)^T)}{\sum_{i=1}^{N} p(z_i=1\|x_i,\theta^{(t)})}$ (6.17)

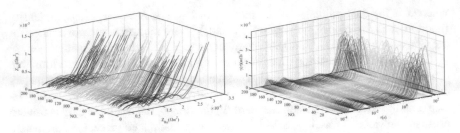

Fig. 6.16 EIS and DRT plots of 200 retired cells

where $a(x_i)$ is the average distance between x_i and other data points in the cluster, and $b(x_i)$ is the average distance from x_i to all other points in the cluster.

The range of $s(x_i)$ is distributed from -1 to 1. Specifically, if $-1 \leq s(x_i) < 0$, the clustering is unreasonable, and if $s(x_i) = -1$, the clustering becomes the worst. If $0 \leq s(x_i) \leq 1$, the clustering is reasonable, and if $s(x_i) = 1$, the clustering becomes the best.

To fully demonstrate the advantages of the soft clustering method, 200 retired cells with different ageing degrees are tested by EIS and DTR described in Sect. 6.2.2, and the results are shown in Fig. 6.16. According to the above soft clustering method using GMM, the 200 cells are soft clustered into five groups, which are labelled C1, C2, C3, C4, and C5. Figure 6.17 shows the clustering results, in which the abscissa is the cell number, and the ordinate is the clustering probability. $P(CX)$ ($X = 1, 2, 3, 4, 5$) is the probability that one sample belongs to groups, and the P is distributed from 0 to 1. Specifically, the green, red, and white represent the weak, strong, and median probability of a battery belonging to a cluster, respectively. It can be observed that some cells can be grouped into different clusters at the same time. These cells with middle colours can be flexibly shared by multiple groups during regrouping, which means that these cells are soft clustered. Furthermore, if a cell with the probability of $0.3 < P < 0.7$ is defined as a cell with soft clustering characteristics, then there are 17 cells belonging to multiple categories that can be flexibly regrouped. It can be inferred that cells with soft clustering characteristics will increase as the number of cells increases, which greatly increases the regrouping flexibility of the large-scale retired Li-ion batteries.

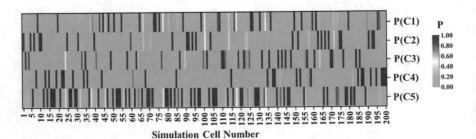

Fig. 6.17 Soft clustering results of the 200 cells using GMM

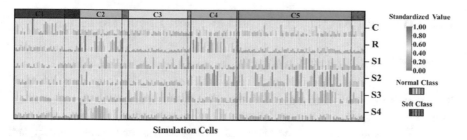

Simulation Cells

Fig. 6.18 Consistency of the sorting criteria within each cluster

Figure 6.18 shows the distribution of the six sorting criteria for each cell after clustering, where the abscissa is the cell label and the ordinate is the normalized value of the six criteria. The five groups are marked with different colours. The cells in the solid colour group do not have soft clustering characteristics, while the cells in the shaded colour group have soft clustering characteristics. It can be observed that the six sorting criteria of the same group of cells all have the good consistency. For example, C1 has a large and consistent capacity, while C2 and C4 have large and consistent internal resistances. It shows that clustering based on the constructed multidimensional criteria can express the key characteristics of the battery, and then the refined regrouping can be carried out. In addition, the SC value calculated according to Eq. (6.12) is 0.478, indicating that the proposed algorithm has a good clustering effect.

It should be noted that the cells in the shaded colour group of Fig. 6.19 are shared by multiple clusters. The value in the grey grid is the number of non-soft clustered cells in each group, and the white grid is the number of soft clustered cells in each group. For example, three cells are shared between C1 and C2, and six cells are shared between C1 and C5. It can be concluded that the proposed clustering method based on multidimensional criteria can improve the accuracy and rationality of battery regrouping, and the soft clustering algorithm based on GMM can improve the flexibility of large-scale battery regrouping.

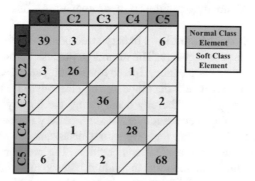

Fig. 6.19 Battery distribution with the soft clustering characteristics

6.4   Material Recycling Method of Spent Li-Ion Batteries

6.4.1   Main Recycling Methods

The material recycling and reuse of spent Li-ion batteries is an important part of the closed-loop of the Li-ion battery full-lifespan cycle, which can realize the recycling of valuable resources, reduce the impact of waste treatment on the environment, and reduce the consumption of natural resources. Therefore, the recycling of spent Li-ion batteries has received widespread attention in recent years. Big data technology plays an increasingly important role in the lifecycle management of Li-ion batteries. The battery recycling traceability comprehensive management system is under construction and improvement in many countries around the world. During the use of the battery, the information of battery maintenance or battery retirement is transmitted to the platform by the vehicle manufacturer. After the battery is retired from EVs, the battery disassembly, echelon utilization, and recycling enterprises will submit the battery-related information. In all processes, the material flow in the full-lifespan cycle of batteries will be submitted to the traceability integrated management platform, and this submitted information forms the big data of batteries. Through the platform, we can understand the information of the waste batteries in a certain area, and guide the establishment of battery recycling enterprises and the application of battery recycling methods.

The recycling process of the spent Li-ion batteries is to separate the useful components in the batteries by using their physical and chemical properties to realize the reuse of resources. The recycling methods of the spent Li-ion batteries based on battery data can be roughly divided into physical and chemical methods, as shown in Fig. 6.20. At present, physical methods are mostly used as the pretreatment before the chemical methods. The physical method includes mechanical separation, heat treatment, mechanochemical treatment, and dissolution treatment. Chemical methods are an effective solution to improve material recycling rate and purity. The chemical treatment process includes acid leaching, biological leaching, solvent extraction, chemical precipitation, and electrochemical treatment. At present, the main chemical recycling methods are pyrometallurgy, hydrometallurgy, bioleaching, and mixed treatment.

Due to the complexity of the Li-ion battery structure, pretreatment is required to achieve the maximum recovery rate. The discharge process before disassembly will reduce the energy of the waste battery, thereby preventing spontaneous combustion. The heat treatment removes and decomposes the electrolyte by thermochemically degrading organic compounds, thereby deactivating the battery. The battery pack includes many large components such as housing, battery management system, or cooling components. Therefore, they were first disassembled manually and then sorted according to size and chemical composition. The mechanical treatment can reduce material volume and separate individual battery materials. To release the positive and negative materials, Li-ion batteries must be crushed, grounded, and then

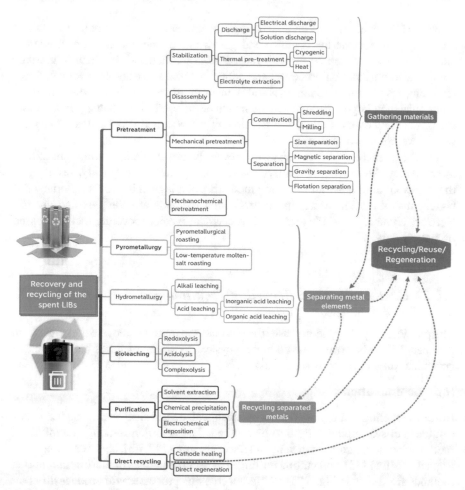

Fig. 6.20 Main recycling methods for the spent Li-ion batteries, reprinted from [40], with permission from Elsevier

sieved. High-energy grinding is used to reduce the particle size and increase the specific surface area, further simplifying the leaching process.

In pyrometallurgical processing, battery components are smelted at high temperatures to obtain a metal alloy composed of metals Cu, Ni, Co, and Fe, and then purified and separated by hydrometallurgy. Mn and Ti are usually not recycled as metals, but oxidized and form slag. In hydrometallurgical recycling, the cathode material is dissolved in acid, and individual metals are separated by solvent extraction. Inorganic acids are used to dissolve metal components during the leaching process. Subsequently, the metal is concentrated and purified by chemical precipitation, ion exchange, or solvent extraction. Compared with pyrometallurgical processes, hydrometallurgy has higher recovery efficiency, lower energy consumption, and lower emissions. However, hydrometallurgical technology has complicated process steps, high consumption of chemical reagents, and environmental pollution. Direct recycling is to recycle the negative and positive electrode materials as an integral part, and thus, they can be directly reused in Li-ion battery manufacturing. Since complicated purification processes and active material synthesis are avoided, direct recycling has economic advantages and is environmental friendly. However, its recycling efficiency largely depends on the health of the used Li-ion battery. In addition, the physical recovery method is a developing method, and it is expected to have a good development prospect.

Owing to there are many ways to recycle battery materials, some are still in their infancy. Moreover, the battery material is updated very quickly, resulting in the corresponding material recycling methods often lag behind the development of battery materials. Therefore, battery recycling methods are complex and diverse. Here are some cases to illustrate several classical battery recycling methods based on battery data.

6.4.2   Case 1: Physical Recycling Technologies

The physical recycling method uses the physical separation to recycle materials from the spent Li-ion batteries. The common physical recycling methods mainly include comminution and physical separation as:

(1)   **Comminution**

Before recycling battery materials by chemical or physical methods, the disassembled cells or modules need to be crushed. The common rotating comminution methods include hammer crushing [41], wet crushing [42, 43], shear crushing [44], impact crushing [45], and cutting milling. The main principles of these comminution methods are shown in Fig. 6.21a. Different crushing processes will produce different sizes and shapes of materials, which will seriously affect the subsequent separation processes [46, 47].

(2)   **Separation**

The physical separation is a commonly used technology to facilitate subsequent material recycling. It uses the physical properties of the mixture (e.g. colour, density, magnetic properties, particle size, and surface physical properties) to separate components from waste Li-ion batteries as much as possible. The commonly used physical separation methods are summarized as follows:

Size separation: Size separation is a common process for the preliminary separation of the crushed spent batteries. It is usually realized by the vibrating screen, and its main principle is shown in Fig. 6.21b. The comminuted mixture can be divided into fine particles (<1 mm) and coarse particles (>1 mm) by size. Generally, the coarse particles mainly consist of plastic, separator, aluminium foil, and copper foil, while the fine particles mainly consist of positive and negative materials. In Ref. [46], the

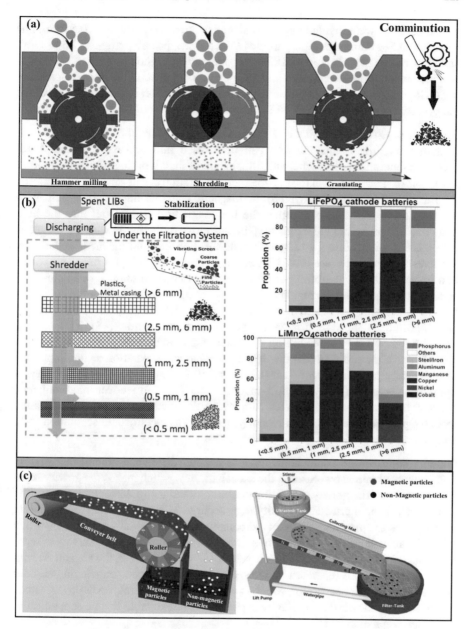

Fig. 6.21 Schematic illustration of the commonly used physical recycling methods. **a** Rotating comminution; **b** size separation; **c** magnetic separation, reprinted from [40], with permission from Elsevier

mixed particles were divided into five categories: ultrafine particles (<0.5 mm), fine particles (0.5–1 mm), medium particles (1–2.5 mm), coarse particles (2.5–6 mm), and ultra-coarse particles (>6 mm). The separation results indicated that 82% Co was obtained in the ultrafine particles and 68% Co was obtained in the fine particles. For the NCM batteries, Co is the most valuable metal, while Ni is abundant. Therefore, the potential recycling value of Ni in cathode materials would exceed that of Co.

Magnetic separation: Magnetic separation is an effective method to separate metals from non-metallic components. It can be divided into dry-magnetic and fluid-magnetic separations. The schematic illustration of their working principle is shown in Fig. 6.21c. The process of the dry-magnetic separation can be described as follows: the mixed particles are conveyed to the magnetic roller by the conveyor belt. With the rotation of the magnetic roller, the non-magnetic particles fall into the non-magnetic collector, and the magnetic particles move to the baffle with the magnetic roller, and then fall into the magnetic collector. For the fluid-magnetic separation, the material and water are mixed based on a certain solid–liquid ratio and stirred at a little speed. The uniformly stirred suspension flows to the magnetic plate at a constant flow rate, and the magnetic particles are tightly adsorbed on the collecting mat under the action of magnetic force, while non-magnetic particles are taken away and collected by the collector under the action of water flow [48]. This method provides a new idea for separating the micromagnetic particles from mixed particles. To solve the problem that the existing mechanical separation technologies can only separate particles larger than 0.075 mm, Ref. [49] proposed a technology of ultrasonic dispersion and waterflow-magnetic separation to recover micromagnetic particles from mixed microparticles. Moreover, other metal and non-metal separation technologies were developed in recent years, such as electrostatic separation [50], eddy current separation [51].

Gravity separation: There is an obvious density difference among the mixture components obtained by crushing the spent batteries, which makes the gravity separation possible. The gravity separation can be achieved using shaker tables, vibrating screens, a fluid of intermediate density, or air separation. Reference [52] used different airflow rates to spray and clean mixed components for separation. It is shown that when the air velocity is 10.2–10.5 m s^{-1}, the smaller diameter polymers were separated, such as Cu and Al; when the air velocity is 10.6–13 m s^{-1}, the Cu and Al with larger diameter were separated directly. Moreover, the falcon centrifugal classifier is an efficient gravity separator, which is widely used to separate cathode and anode materials [53]. However, the gravity separation is not suitable for the separation of fine electrode materials [54].

Flotation separation: Flotation separation is an efficient process to separate fine particles based on the difference in surface hydrophilicity of mixtures [55]. For example, the cathode materials in spent Li-ion batteries are hydrophilic, while the anode materials are hydrophobic [56], which provides a basis for their flotation separation. However, the electrode material of spent Li-ion batteries is wrapped by organic matter, further reducing the difference in hydrophilicity between the positive electrode and negative electrode materials. Therefore, surface modification is a necessary step to improve flotation efficiency. Some surface modification methods, such as aerobic coasting [56], anaerobic pyrolysis [57], Fenton high-order oxidation [58], mechanical grinding [59], and cryogenic grinding [60], are applied to remove organic matters on the surface of the electrode materials. Owing to the low recycling rate of resources and serious environmental pollution, the aerobic coasting is less recommended. Although the Fenton high-order oxidation can remove organic matter from electrode materials, Fe^{2+} is introduced into the solution to activate the reaction [61]. Fe^{2+} remains on the surface of electrode materials, which complicates the subsequent metallurgical process. The mechanical grinding can remove some organic matter on the surface of electrode materials. However, the organic binder and electrolyte remain on the surface of electrode particles, resulting in a low recycling rate of cathode materials. Reference [62] proved that pyrolysis-assisted surface modification is an effective method to improve flotation efficiency.

It should be noted that the above physical separation methods may be combined to improve the recycling rate. First, the mixture can be separated by size after the battery is crushed. Then, the steel shell and ferromagnetic material can be removed by magnetic separation. Third, the separator and packaging can be recovered by density or electrostatic separations. Furthermore, the separation of plastics is realized by density separation. In recent years, some innovative methods have been proposed for physical recycling. Reference [63] reports a method of delaminating Li-ion battery electrode by using high-power ultrasonic generator. The adhesion between active material and collector can be quickly broken by the high-power ultrasonic. When the electrode is directly under the high-power ultrasonic generator, the stratification time of the electrode is less than 10 s, and the recycling efficiency of the proposed method is 100 times higher than that of the traditional methods. Reference [64] proposed a battery physical recycling method in which the battery is fully charged and then is placed it in water. This method can easily separate the negative electrode material and negative collector to obtain lithium salt. The physical recycling methods have the advantages of short process and being environment-friendly. The development of innovative physical recycling methods has great significance and bright prospects.

6.4.3 Case 2: Chemical Recycling Technologies

The current chemical recycling technologies of the spent Li-ion batteries mainly include pyrometallurgical, hydrometallurgical, and biometallurgical technologies. The pyrometallurgical technology generally does not require pretreatment, and the technical complexity is low. However, lithium and aluminium are easily discharged with the slag and cannot be recovered, and the metallurgical process has high-energy consumption and serious environmental pollution, which does not meet the requirements of environmentally friendly industries. Biometallurgy is the use of microorganisms to secrete inorganic or organic acids to recover metal substances in spent

Li-ion batteries. This technology has the advantages of low energy consumption and low secondary pollution, but the microbial cultivation conditions are harsh and the leaching cycle is long. In comparison, hydrometallurgical technology has attracted more and more attention due to its advantages of high recycling efficiency, low cost, low energy consumption, and low secondary pollution. The mature technology of pyrometallurgy and hydrometallurgy is summarized as follows:

(1) Pyrometallurgical technologies

Conventional pyrometallurgical technologies can be classified into pyrolysis and reduction roasting. The basic principle of pyrolysis is that the electrode materials will be converted into relatively stable oxidation or metal states under a high-temperature environment. This method is generally used for the recycling of cathode materials. Reference [65] revealed a unique phase transition behaviour of $Li_{1/3}Ni_{1/3}Co_{1/3}Mn_{1/3}O_2$ cathode during heating: the initial layer structure first transformed into an Li_2O_4-type from 236 to 350 °C, and then to a M_3O_4-type spinel from 350 to 441 °C. Reference [66] indicated that the content of Ni, Co, and Mn in NCM cathode materials significantly affects the structural changes during heating, and the more Ni and less Co and Mn, the lower the temperature of the phase transition, as shown in Fig. 6.22a. Reference [67] proposed a new method for predicting the thermodynamics of thermal degradation of the cathode materials of Li-ion batteries. In summary, cobalt, nickel, copper, and other metals are melted and recovered as alloys in pyrolysis, which needs the subsequent treatment. In addition, lithium and other components will be discarded in the form of slag and gas at high temperatures, resulting in the loss of valuable metals. Furthermore, the high-energy consumption and toxic gas emission are the other defects of the pyrolysis method for the recycling of retired Li-ion batteries.

In the reduction roasting method, coke, carbon monoxide, and active metals are used as reducing agents to reduce metals from their compounds. Generally, the anode materials of battery are usually used as high-temperature reducing agents for the recycling of cathode materials. In Ref. [68], the NCM cathode materials were reduced to Li_2CO_3, MnO, NiO, Ni, and Co by calcining with coke as a reducing agent at 650 °C for 30 min. In Ref. [69], the LCO cathode material was reduced from the crystal structure of the cathode material to form Li_2CO_3 at high temperature. In Ref. [70], the cathode and anode materials of LCO were calcined together, and the following coupling reactions have occurred:

$$\begin{cases} 4LiCoO_2 + C \rightarrow 2Li_2O + CO_2 + 4CoO \\ 4LiCoO_2 + 2C \rightarrow 2Li_2O + 2CO + 4CoO \\ Li_2O + CO + CoO \rightarrow Co + Li_2CO_3 \\ Li_2O + C + 2CoO \rightarrow 2Co + Li_2CO_3 \end{cases} \quad (6.21)$$

From the point of view of crystal structure, the stronger attraction of graphite to oxygen than lithium and cobalt would make oxygen octahedrons inside lithium cobalt oxide unstable to break down. The whole process is illustrated in Fig. 6.22b.

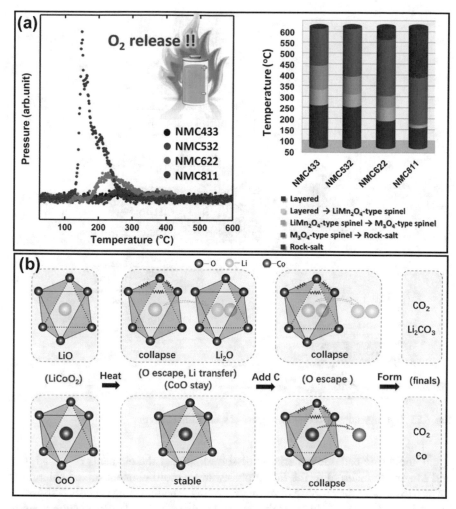

Fig. 6.22 Cases of pyrometallurgy methods. **a** Temperature region of the phase transitions for NMC; **b** collapsing model of recycling metals from LiCoO$_2$ by roasting, reprinted from [40], with permission from Elsevier

In addition, binders, separators, electrolytes, aluminium shells, plastics, and by-products of chemical plants (e.g. sulphur-containing tail gas, and slag) can also be the reducing agents for cathode materials. However, with the development of cobalt-free electrode materials in Li-ion batteries, high-temperature roasting technology is facing challenges.

(2)   **Hydrometallurgical technologies**

Hydrometallurgy is a common recycling method, which mainly uses the acid or alkali systems as the leaching agent. Under the combined action of the reducing

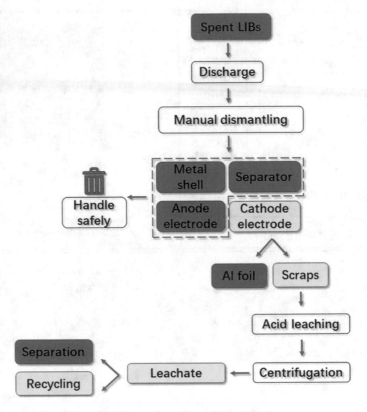

Fig. 6.23 Hydrometallurgy recycling process of spent Li-ion batteries

agent, the waste cathode material is dissolved, so that the elements of Li, Ni, Co, and Mn are transferred to the liquid phase, and a multi-element mixed solution is formed to achieve the purpose of further recycling. The recycling process is shown in Fig. 6.23, and it can be observed that it has the characteristics of low environmental pollution and high recovery efficiency. The core step of hydrometallurgy recycling is the leaching process, which is mainly divided into two types: acid leaching and alkali leaching. Among them, acid leaching is the more commonly used method. In the acid leaching process, it is usually necessary to use a reducing agent as auxiliary material to reduce the high-valent transition metal elements in the waste cathode material to a low-valent state, thereby accelerating the leaching process. The acid leaching agents mainly include sulphuric acid, hydrochloric acid, nitric acid, phosphoric acid, and hydrofluoric acid, and the alkali leaching agents mainly include ammonia and ammonium sulphate. The main reducing agents include hydrogen peroxide, sodium sulphite, sodium bisulphite, sodium thiosulphate, ammonium chloride, etc. From the literature, the acid leaching method has absolute advantages in the process of cathode waste from solid state to ionic state. Generally, organic acid or inorganic acid is used as leaching agent, the M–O (M=Ni, Co, Mn) bond in the cathode material is destroyed by H+ structure, and then the metal is leached in ionic state. Inorganic acid has strong acidity, which can dissolve most valuable metals into ionic state and enter the solution, but its corrosion to the reaction vessel is also very serious; organic acid has weak acidity, has special spatial structure and binding site, and is widely used because of its advantages of the high recovery rate of valuable metals, low pollution, and easy control Pan application. The following two leaching methods are briefly introduced.

Inorganic acid leaching methods: The inorganic acids commonly used for solvent leaching of valuable metals in electrode waste generally include strong acids such as HCl, H_2SO_4, HNO_3, and H_3PO_4, among which HCl has the highest leaching efficiency. Under the optimum leaching conditions of HCI, the leaching rates of Ni, Co, Mn, and Li can reach more than 99%. However, HCl is volatile and easy to produce harmful gas, which requires special anti-corrosion equipment. The recycling process would cause great pollution to the environment, resulting in increased recycling costs.

In recent years, inorganic acid leaching has been widely concerned for its high leaching efficiency, high selectivity of leaching agent, and mature technology. Some representative leaching results of different cathode materials in different inorganic acids are listed in Table 6.7. Generally, the leaching effect is affected by the concentration of leaching agent, the amount of reducing agent, leaching temperature, leaching time, and solid–liquid ratio. Reference [71] proposed a two-step leaching method to extract valuable metals selectively from $LiNi_xCo_yMn_{1-x-y}O_2$ cathode materials: First, Ni, Co, and Li were leached from lixivium either as complexes or metallic ion by employing ammoniacal solution as the leaching agent and sodium sulphite as reductant. Second, manganese was deposited from Mn_3O_4 to $(NH_4)_2Mn(SO_3)_2 \cdot H_2O$ as sodium sulphite was added. The loose and porous Mn_3O_4 is more favourable for ion diffusion and leaching reaction, as shown in Fig. 6.24.

Moreover, the use of alkaline systems in hydrometallurgy recycling can also achieve an excellent leaching effect. For example, under the combined action of the leaching agent of ammonia and ammonium carbonate, and the reducing agent of ammonium sulphite, Co and Cu can be selectively leached and recovered from waste $LiMn_2O_4$ and ternary materials. Through the above analysis, it can be concluded that a suitable leaching system can successfully achieve the effective leaching and recycling of metal elements in the cathode materials of the waste Li-ion batteries. The recycling product is a mixed solution of multiple elements such as Li, Ni, Co, and Mn. In the subsequent processing steps, on the one hand, the metal salt compounds can be extracted from the leachate by selective separation and used as industrial raw materials; at the same time, the leachate can be directly used as the raw material of the electrode material regeneration process to improve the recovery of valuable metals.

From the literature, the leaching efficiency of the inorganic acid is very high. However, the wastewater, waste residue, and harmful gases (e.g. SO_2, SO_3, NO_x) will be produced in the leaching process, which poses a great threat to the ecological environment and human health. Therefore, green environmental protection, high efficiency, low-cost leaching agent leaching method are the future trend and direction.

Table 6.7 Summary of leaching results for the spent Li-ion batteries in inorganic different acids

Cathode materials	leaching agents	Leaching results	References
NCM	H_2SO_4	92% of Li, 92% of Ni, 68% of Co and 34.8% of Mn	[72]
NCM	$H_2SO_4 + H_2O_2$	99.7% of Li, Ni, Co, and Mn	[73]
NCM	$H_2SO_4 + H_2O_2$	98% of Ni, 99% of Co, and 84% of Mn	[74]
NCM	$H_2SO_4 + H_2O_2$	99.8% of Li, 96.46% of Co	[75]
NCM	$H_2SO_4 + Na_2S_2O_5$	85% of Li, 90% of Ni, Co and Mn	[76]
NCM	$H_2SO_4 + NH_4Cl$	99.1% of Li, 97.4% of Ni, 97.5% of Co, and 97.3% of Mn	[77]
NCM	$HNO_3 + HCl$	71% of Li, 33% of Ni, 34% of Co and 40% of Mn	[78]
LCO	H_2SO_4	99% of Li, 99% of Co	[79]
LCO	$H_2SO_4 + H_2O_2$	95% of Li, 80% of Co	[80]
LCO	HCl	99% of Li, 99% of Co	[81]
LCO	$HNO_3 + H_2O_2$	95% of Li, 95% of Co	[82]
LCO	$H_2SO_4 + Na_2S_2O_3$	99.71% of Li, 99.95% of Co	[83]
LCO	$H_2SO_4 + glucose$	92% of Li, 88% of Co	[84]
LCO	$H_3PO_4 + H_2O_2$	99% of Li, 99% of Co	[85]
LFP	$Na_2S_2O_8$	99% of Li	[86]
LFP	H_3PO_4	97.67% of Fe, and 94.29% of Li	[87]
LFP	$H_2SO_4 + H_2O_2$	96.85% of Li	[88]
LFP + LMO	$HCl + H_2O_2$	80.93% of Li, 85.40% of Fe and 81.02% of Mn	[89]

Organic acid leaching methods: Since it is inevitable that volatile and toxic gases will be generated during the use of inorganic acids, which are harmful to human health and pollute the environment, and the current research focuses on the use of acid system recovery methods is also on some natural organic acids in recent years. Although the acidity of organic acids is lower than that of inorganic acids, some organic acids still show quite a good leaching rate in the leaching process, which is mainly due to the formation of complexes between organic acid radical ions and valuable metal cations [90]. According to the different leaching mechanisms, organic acids are divided into chelating organic acids, reducing organic acids, precipitating organic acids, and other organic acids [91]. The common chelating organic acids are citric acid, malic acid, platinum succinate, and aspartic acid, and the common reducing organic acids are ascorbic acid ($C_6H_8O_6$) and lactic acid ($C_3H_6O_3$). The typical precipitating organic acid is oxalic acid ($H_2C_2O_4$). In Ref. [92], $C_3H_6O_3$ was chosen as the leaching and chelating agent to recycle the cathode materials from spent Li-ion batteries. Table 6.8 lists some leaching effects of different cathode materials in different organic acids. After leaching valuable metals with organic acids, the leaching solution or recovered products are used to replace the raw materials for the

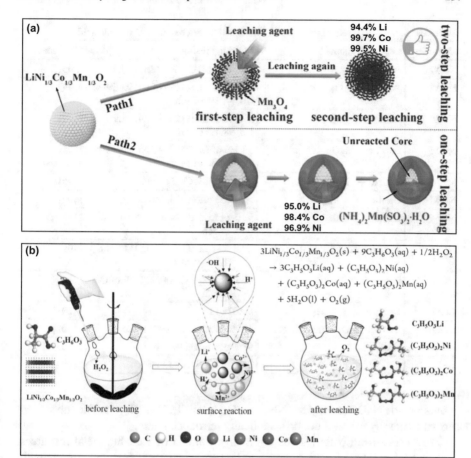

Fig. 6.24 Process of a two-step leaching method, reprinted from [40], with permission from Elsevier

resynthesis of electrode materials, to realize the closed-loop recycling of the spent Li-ion batteries.

The waste cathode material recovered by the above method is usually a mixed system of several different metal ions. If a certain single metal ion is required in the subsequent process, the next step of separation operation is required, such as solvent extraction, chemical precipitation. The solvent extraction uses a two-phase system (usually an organic phase and an aqueous phase) to achieve separation through the uneven distribution of different ions in the two phases. The separation mechanism of chemical precipitation is the different solubility of metal compounds at a certain pH. In general, the solubility of transition metal hydroxides and oxalates is much lower compared to the corresponding lithium compounds. Thus, different metal ions can be used for cascaded precipitation separation at different pH values to efficiently separate metal ions such as Ni, Co, Mn, and Li. The commonly used precipitation

Table 6.8 Summary of leaching results for the spent Li-ion batteries in organic acid

Cathodes	Leaching agents	Leaching results	References
NCM	Acetic + H_2O_2	>98% of Li, Co, Ni, Mn	[93]
NCM	Acetic + H_2O_2	99.9% of Li, 92.6% of Ni, 93.6% of Co, 96.3% of Mn	[94]
NCM	Lactic + H_2O_2	97.7% of Li, 98.2% of Ni, 98.9% of Co, 98.4% of Mn	[92]
NCM	Formic + H_2O_2	98.2% of Li, 99.9% of Ni, 99.9% of Co, 99.9% of Mn	[95]
NCM	Maleic + H_2O_2	> 98% of Li, Co, Ni, Mn	[96]
NCM	Oxalic	> 98.5% of Ni, Co, Mn	[97]
NCM	Citric + D-glucose	99% of Li, 91% of Ni, 92% of Co, 94% of Mn	[98]
NCM	Trichloroacetic acid + H_2O_2	89.8% of Li, 93% of Ni, 91.8% of Co, 91.8% of Mn	[99]
NCM	Citrus fruit juice	100% of Li, 98% of Ni, 94% of Co, 99% of Mn	[100]
LCO	Citric acid + H_2O_2	99.07 of Li	[101]
LCO	Tartaric acid + ascorbic acid	100% of Li, and >90% of Co	[102]
LCO	Oxalic acid	98% of Li, and 97% of Co	[103]
LFP	Oxalic	99% of Li, 94% of Fe	[104]

agents include $NaOH$, $H_2C_2O_4$, $(NH_4)_2C_2O_4$, Na_2CO_3, Na_3PO_4, etc. The subsequent processes are diverse and not be specifically introduced here.

The waste Li-ion batteries contain a large number of valuable metal resources, such as Ni, Co, Mn, Li, which have good prospects for recycling and are gradually being valued. At the same time, the recycling of large amounts used Li-ion batteries poses new challenges to environmental protection and sustainable use of resources and puts tremendous pressure on the development of appropriate recycling technologies. The pyrometallurgical recycling process has been widely studied due to its short process and high efficiency. The hydrometallurgical recycling process has gradually become a research hotspot due to its good selectivity to valuable metals and mild reaction conditions. It is relatively mature and worthy of promotion. The recycling mechanism of hydrometallurgy should be studied continuously, and the battery industry chain, the related recycling processes, and equipment should be perfected. Under the continuous improvement of battery recycling policies, a comprehensive utilization system should be established, a market recycling system should be improved. More importantly, cascade utilization and battery dismantling and recycling should be more effectively integrated and developed, and the waste Li-ion battery companies and other energy companies should coexist and develop in harmony.

6.5 Summary

This chapter describes the data science-based battery reutilization management for retired Li-ion batteries. First, the echelon utilization and material recycling of retired Li-ion batteries are briefly introduced. They can maximize the full-lifespan value of Li-ion batteries and alleviate the pressure of lithium, cobalt, manganese, and other resources, which is of great significance to the sustainable development of Li-ion batteries. Second, aiming at the sorting problem in the echelon utilization of large-scale Li-ion batteries, a data-based sorting criterion is constructed, the test data is used to train the neural network model, and the machine learning algorithm is used to complete the rapid sorting of large-scale Li-ion batteries. In addition, two cases are used to verify the proposed fast sorting method: one case is used to describe the fast estimation of battery capacity and internal resistance using partial charging data based on the proposed method; the other case describes a fast estimation method of battery capacity and other important characteristic parameters using EIS, and the experimental results show that the speed of obtaining battery capacity by the proposed method is 10 times higher than that by the traditional method. Third, aiming at the problem of battery regrouping in echelon utilization, a hard-clustering method based on K-means and a soft clustering method based on the GMM algorithm are proposed. Finally, the material recycling methods are comprehensively summarized and discussed. The establishment of a traceability management platform for material and information flows of Li-ion batteries based on big data in the full-lifespan cycle can greatly facilitate the echelon utilization and material recycling of Li-ion batteries.

References

1. Pellow MA, Ambrose H, Mulvaney D, Betita R, Shaw S (2020) Research gaps in environmental life cycle assessments of lithium ion batteries for grid-scale stationary energy storage systems: end-of-life options and other issues. SM&T 23:e00120
2. Wang W, Wu Y (2017) An overview of recycling and treatment of spent LiFePO$_4$ batteries in China. Resour Conserv Recycl 127:233–243
3. Yang Y, Xu S, He Y (2017) Lithium recycling and cathode material regeneration from acid leach liquor of spent lithium-ion battery via facile co-extraction and co-precipitation processes. Waste Manage 64:219–227
4. Diekmann J, Hanisch C, Froböse L, Schälicke G, Loellhoeffel T, Fölster A-S, Kwade A (2016) Ecological recycling of lithium-ion batteries from electric vehicles with focus on mechanical processes. J Electrochem Soc 164(1):A6184
5. Winslow KM, Laux SJ, Townsend TG (2018) A review on the growing concern and potential management strategies of waste lithium-ion batteries. Resour Conserv Recycl 129:263–277
6. Meshram P, Mishra A, Sahu R (2020) Environmental impact of spent lithium ion batteries and green recycling perspectives by organic acids—a review. Chemosphere 242:125291
7. Mayyas A, Steward D, Mann M (2019) The case for recycling: Overview and challenges in the material supply chain for automotive li-ion batteries. SM&T 19:e00087
8. Richa K, Babbitt CW, Gaustad G, Wang X (2014) A future perspective on lithium-ion battery waste flows from electric vehicles. Resour Conserv Recycl 83:63–76
9. Zeng X, Li J, Liu L (2015) Solving spent lithium-ion battery problems in China: opportunities and challenges. Renew Sust Energ Rev 52:1759–1767

10. Huang B, Pan Z, Su X, An L (2018) Recycling of lithium-ion batteries: recent advances and perspectives. J Power Sources 399:274–286
11. Lai X, Huang Y, Deng C, Gu H, Han X, Zheng Y, Ouyang M (2021) Sorting, regrouping, and echelon utilization of the large-scale retired lithium batteries: A critical review. Renew Sust Energ Rev 146:111162
12. Liao Q, Mu M, Zhao S, Zhang L, Jiang T, Ye J, Shen X, Zhou G (2017) Performance assessment and classification of retired lithium ion battery from electric vehicles for energy storage. Int J Hydrog Energy 42(30):18817–18823
13. Schneider E, Kindlein W Jr, Souza S, Malfatti C (2009) Assessment and reuse of secondary batteries cells. J Power Sources 189(2):1264–1269
14. Jiang Y, Jiang J, Zhang C, Zhang W, Gao Y, Guo Q (2017) Recognition of battery aging variations for LiFePO4 batteries in 2nd use applications combining incremental capacity analysis and statistical approaches. J Power Sources 360:180–188
15. Jiang Y, Jiang J, Zhang C, Zhang W, Gao Y, Li N (2018) State of health estimation of second-life LiFePO$_4$ batteries for energy storage applications. J Clean Prod 205:754–762
16. Rohr S, Müller S, Baumann M, Kerler M, Ebert F, Kaden D, Lienkamp M (2017) Quantifying uncertainties in reusing lithium-ion batteries from electric vehicles. Procedia Manuf 8:603–610
17. Lai X, Qiao D, Zheng Y, Ouyang M, Han X, Zhou L (2019) A rapid screening and regrouping approach based on neural networks for large-scale retired lithium-ion cells in second-use applications. J Clean Prod 213:776–791
18. Li X, Zhang L, Liu Y, Pan A, Liao Q, Yang X (2020) A fast classification method of retired electric vehicle battery modules and their energy storage application in photovoltaic generation. Int J Energy Res 44(3):2337–2344
19. Barai A, Chouchelamane GH, Guo Y, Mcgordon A, Jennings P (2015) A study on the impact of lithium-ion cell relaxation on electrochemical impedance spectroscopy. J Power Sources 280:74–80
20. Itagaki M, Honda K, Hoshi Y, Shitanda I (2015) In-situ EIS to determine impedance spectra of lithium-ion rechargeable batteries during charge and discharge cycle. J Electroanal Chem 737:78–84
21. Babaeiyazdi I, Rezaei-Zare A, Shokrzadeh S (2021) State of charge prediction of EV Li-ion batteries using EIS: a machine learning approach. Energy 223:120116
22. Kurc B, Pigłowska M (2021) An influence of temperature on the lithium ions behavior for starch-based carbon compared to graphene anode for LIBs by the electrochemical impedance spectroscopy (EIS). J Power Sources 485:229323
23. Fan B, Guan Z, Wang H, Wu L, Li W, Zhang S, Xue B (2021) Electrochemical processes in all-solid-state Li-S batteries studied by electrochemical impedance spectroscopy. Solid State Ion 368:115680
24. Waluś S, Barchasz C, Bouchet R, Alloin F (2020) Electrochemical impedance spectroscopy study of lithium–sulfur batteries: useful technique to reveal the Li/S electrochemical mechanism. Electrochim Acta 359:136944
25. Zhou X, Huang J, Pan Z, Ouyang M (2019) Impedance characterization of lithium-ion batteries aging under high-temperature cycling: importance of electrolyte-phase diffusion. J Power Sources 426:216–222
26. Choi W, Shin H-C, Kim JM, Choi J-Y, Yoon W-S (2020) Modeling and applications of electrochemical impedance spectroscopy (EIS) for lithium-ion batteries. Electrochim Acta 11(1):1–13
27. Wang X, Wei X, Zhu J, Dai H, Zheng Y, Xu X, Chen Q (2021) A review of modeling, acquisition, and application of lithium-ion battery impedance for onboard battery management. ETransportation 7:100093
28. Celebi ME, Kingravi HA, Vela PA (2013) A comparative study of efficient initialization methods for the k-means clustering algorithm. Expert Syst Appl 40(1):200–210
29. Lai X, Deng C, Li J, Zhu Z, Han X, Zheng Y (2021) Rapid sorting and regrouping of retired lithium-ion battery modules for echelon utilization based on partial charging curves. IEEE Trans Veh Technol 70(2):1246–1254

30. Ghassabeh YA, Rudzicz F (2021) Modified subspace constrained mean shift algorithm. J Classif 38(1):27–43
31. Ramalakshmi K, Raghavan VS (2021) Kernalized average entropy and density based spatial clustering with noise. J Ambient Intell Humaniz Comput 12(3):3937–3947
32. Chaurasia N, Tapaswi S, Dhar J (2018) A resource efficient expectation maximization clustering approach for cloud. Comput J 61(1):95–104
33. Al-Dabooni S, Wunsch D (2018) Model order reduction based on agglomerative hierarchical clustering. IEEE Trans Neural Netw Learn Syst. 30(6):1881–1895
34. Choubin B, Moradi E, Golshan M, Adamowski J, Sajedi-Hosseini F, Mosavi A (2019) An ensemble prediction of flood susceptibility using multivariate discriminant analysis, classification and regression trees, and support vector machines. Sci Total Environ 651:2087–2096
35. De Carvalho FDA, Lechevallier Y, De Melo FM (2012) Partitioning hard clustering algorithms based on multiple dissimilarity matrices. Pattern Recognit 45(1):447–464
36. Ferreira MR, De Carvalho FDA, Simões EC (2016) Kernel-based hard clustering methods with kernelization of the metric and automatic weighting of the variables. Pattern Recognit 51:310–321
37. Wang J, Jiang J (2021) Unsupervised deep clustering via adaptive GMM modeling and optimization. Neurocomputing 433:199–211
38. Asheri H, Hosseini R, Araabi BN (2021) A new EM algorithm for flexibly tied GMMs with large number of components. Pattern Recognit 114:107836
39. Lin Z, Wen F, Ding Y, Xue Y (2017) Data-driven coherency identification for generators based on spectral clustering. IEEE Trans Industr Inform. 14(3):1275–1285
40. Lai X, Huang Y, Gu H, Deng C, Han X, Feng X, Zheng Y (2021) Turning waste into wealth: A systematic review on echelon utilization and material recovery of the retired lithium-ion batteries. Energy Storage Mater
41. Granata G, Pagnanelli F, Moscardini E, Takacova Z, Havlik T, Toro L (2012) Simultaneous recycling of nickel metal hydride, lithium ion and primary lithium batteries: accomplishment of European Guidelines by optimizing mechanical pre-treatment and solvent extraction operations. J Power Sources 212:205–211
42. Zhang T, He Y, Ge L, Fu R, Zhang X, Huang Y (2013) Characteristics of wet and dry crushing methods in the recycling process of spent lithium-ion batteries. J Power Sources 240:766–771
43. Wang F, Sun R, Xu J, Chen Z, Kang M (2016) Recovery of cobalt from spent lithium ion batteries using sulphuric acid leaching followed by solid–liquid separation and solvent extraction. RSC Adv 6(88):85303–85311
44. Zhang T, He Y, Wang F, Ge L, Zhu X, Li H (2014) Chemical and process mineralogical characterizations of spent lithium-ion batteries: an approach by multi-analytical techniques. Waste Manage 34(6):1051–1058
45. Ruffino B, Zanetti M, Marini P (2011) A mechanical pre-treatment process for the valorization of useful fractions from spent batteries. Resources, Resour Conserv Recycl. 55(3):309–315
46. Wang X, Gaustad G, Babbitt CW (2016) Targeting high value metals in lithium-ion battery recycling via shredding and size-based separation. Waste Manage 51:204–213
47. Sommerville R, Shaw-Stewart J, Goodship V, Rowson N, Kendrick E (2020) A review of physical processes used in the safe recycling of lithium ion batteries. SM&T e00197
48. Qiu R, Huang Z, Zheng J, Song Q, Ruan J, Tang Y, Qiu R (2021) Energy models and the process of fluid-magnetic separation for recovering cobalt micro-particles from vacuum reduction products of spent lithium ion batteries. J Clean Prod 279:123230
49. Huang Z, Lin M, Qiu R, Zhu J, Ruan J, Qiu R (2021) A novel technology of recovering magnetic micro particles from spent lithium-ion batteries by ultrasonic dispersion and waterflow-magnetic separation. Resour Conserv Recycl 164:105172
50. Silveira A, Santana M, Tanabe E, Bertuol D (2017) Recovery of valuable materials from spent lithium ion batteries using electrostatic separation. Int J Miner Process 169:91–98
51. Bi H, Zhu H, Zu L, Gao Y, Gao S, Wu Z (2019) Eddy current separation for recovering aluminium and lithium-iron phosphate components of spent lithium-iron phosphate batteries. Waste Manag Res. 37(12):1217–1228

52. Bertuol DA, Toniasso C, Jiménez BM, Meili L, Dotto GL, Tanabe EH, Aguiar ML (2015) Application of spouted bed elutriation in the recycling of lithium ion batteries. J Power Sources 275:627–632

53. Zhang Y, He Y, Zhang T, Zhu X, Feng Y, Zhang G, Bai X (2018) Application of Falcon centrifuge in the recycling of electrode materials from spent lithium ion batteries. J Clean Prod 202:736–747

54. Zhu X-N, Tao Y-J, He Y-Q, Zhang Y, Sun Q-X (2018) Pre-concentration of graphite and LiCoO$_2$ in spent lithium-ion batteries using enhanced gravity concentrator. Physicochem Probl Mineral Process 54

55. Zhu X-N, Nie C-C, Zhang H, Lyu X-J, Qiu J, Li L (2019) Recovery of metals in waste printed circuit boards by flotation technology with soap collector prepared by waste oil through saponification. Waste Manage 89:21–26

56. Wang F, Zhang T, He Y, Zhao Y, Wang S, Zhang G, Zhang Y, Feng Y (2018) Recovery of valuable materials from spent lithium-ion batteries by mechanical separation and thermal treatment. J Clean Prod 185:646–652

57. Zhang G, He Y, Feng Y, Wang H, Zhu X (2018) Pyrolysis-ultrasonic-assisted flotation technology for recovering graphite and LiCoO2 from spent lithium-ion batteries. ACS Sustain Chem Eng 6(8):10896–10904

58. He Y, Zhang T, Wang F, Zhang G, Zhang W, Wang J (2017) Recovery of LiCoO$_2$ and graphite from spent lithium-ion batteries by Fenton reagent-assisted flotation. J Clean Prod 143:319–325

59. Yu J, He Y, Ge Z, Li H, Xie W, Wang S (2018) A promising physical method for recovery of LiCoO$_2$ and graphite from spent lithium-ion batteries: grinding flotation. Sep Purif Technol 190:45–52

60. Liu J, Wang H, Hu T, Bai X, Wang S, Xie W, Hao J, He Y (2020) Recovery of LiCoO$_2$ and graphite from spent lithium-ion batteries by cryogenic grinding and froth flotation. Minerals Eng 148:106223

61. Yu J, He Y, Li H, Xie W, Zhang T (2017) Effect of the secondary product of semi-solid phase Fenton on the flotability of electrode material from spent lithium-ion battery. Powder Technol 315:139–146

62. Zhang G, Du Z, He Y, Wang H, Xie W, Zhang T (2019) A sustainable process for the recovery of anode and cathode materials derived from spent lithium-ion batteries. 11(8):2363

63. Lei C, Aldous I, Hartley JM, Thompson DL, Scott S, Hanson R, Anderson PA, Kendrick E, Sommerville R, Ryder KS (2021) Lithium ion battery recycling using high-intensity ultrasonication. Green Chem

64. Zhao Y, Kang Y, Fan M, Li T, Wozny J, Zhou Y, Wang X, Chueh Y-L, Liang Z, Zhou G (2022) Precise separation of spent lithium-ion cells in water without discharging for recycling. Energy Stor. Mater. 45:1092–1099

65. Nam K-W, Yoon W-S, Yang X-Q (2009) Structural changes and thermal stability of charged LiNi$_1$/3Co$_1$/3Mn$_1$/3O$_2$ cathode material for Li-ion batteries studied by time-resolved XRD. J Power Sources 189(1):515–518

66. Bak S-M, Hu E, Zhou Y, Yu X, Senanayake SD, Cho S-J, Kim K-B, Chung KY, Yang X-Q, Nam K-W (2014) Structural changes and thermal stability of charged LiNi$_x$ Mn$_y$ Co$_z$ O$_2$ cathode materials studied by combined in situ time-resolved XRD and mass spectroscopy. ACS Appl Mater Interfaces 6(24):22594–22601

67. Wang L, Maxisch T, Ceder G (2007) A first-principles approach to studying the thermal stability of oxide cathode materials. Chem Mater 19(3):543–552

68. Liu P, Xiao L, Tang Y, Chen Y, Ye L, Zhu Y (2019) Study on the reduction roasting of spent LiNi$_x$ Co$_y$ Mn$_z$ O$_2$ lithium-ion battery cathode materials. J Therm Anal Calorim 136(3):1323–1332

69. Vishvakarma S, Dhawan N (2019) Recovery of cobalt and lithium values from discarded Li-ion batteries. J Sustain Metall 5(2):204–209

70. Mao J, Li J, Xu Z (2018) Coupling reactions and collapsing model in the roasting process of recycling metals from LiCoO$_2$ batteries. J Clean Prod 205:923–929

71. Meng K, Cao Y, Zhang B, Ou X, Li D-M, Zhang J-F, Ji X (2019) Comparison of the ammoniacal leaching behavior of layered $LiNi_xCo_yMn_{1-x-y}O_2$ (x = 1/3, 0.5, 0.8) Cathode Materials. ACS Sustain Chem Eng 7(8):7750–7759

72. Sattar R, Ilyas S, Bhatti HN, Ghaffar A (2019) Resource recovery of critically-rare metals by hydrometallurgical recycling of spent lithium ion batteries. Sep Purif Technol 209:725–733

73. He L-P, Sun S-Y, Song X-F, Yu J-G (2017) Leaching process for recovering valuable metals from the $LiNi_1/3Co_1/3Mn_1/3O_2$ cathode of lithium-ion batteries. Waste Manage 64:171–181

74. Yang Y, Huang G, Xu S, He Y, Liu X (2016) Thermal treatment process for the recovery of valuable metals from spent lithium-ion batteries. Hydrometallurgy 165:390–396

75. Golmohammadzadeh R, Rashchi F, Vahidi E (2017) Recovery of lithium and cobalt from spent lithium-ion batteries using organic acids: Process optimization and kinetic aspects. Waste Manage (Oxford) 64:244–254

76. Vieceli N, Nogueira CA, Guimarães C, Pereira MF, Durão FO, Margarido F (2018) Hydrometallurgical recycling of lithium-ion batteries by reductive leaching with sodium metabisulphite. Waste Manage 71:350–361

77. Lv W, Wang Z, Cao H, Zheng X, Jin W, Zhang Y, Sun Z (2018) A sustainable process for metal recycling from spent lithium-ion batteries using ammonium chloride. Waste Manage 79:545–553

78. Billy E, Joulié M, Laucournet R, Boulineau A, De Vito E, Meyer D (2018) Dissolution mechanisms of $LiNi_1/3Mn_1/3Co_1/3O_2$ positive electrode material from lithium-ion batteries in acid solution. ACS Appl Mater Interfaces 10(19):16424–16435

79. Nan J, Han D, Zuo X (2005) Recovery of metal values from spent lithium-ion batteries with chemical deposition and solvent extraction. J Power Sources 152:278–284

80. Dorella G, Mansur MB (2007) A study of the separation of cobalt from spent Li-ion battery residues. J Power Sources 170(1):210–215

81. Zhang P, Yokoyama T, Itabashi O, Suzuki TM, Inoue K (1998) Hydrometallurgical process for recovery of metal values from spent lithium-ion secondary batteries. Hydrometallurgy 47(2–3):259–271

82. Lee CK, Rhee K-I (2002) Preparation of LiCoO2 from spent lithium-ion batteries. J Power Sources 109(1):17–21

83. Wang J, Chen M, Chen H, Luo T, Xu Z (2012) Leaching study of spent Li-ion batteries. Procedia Environ. Sci. Eng. Manag. 16:443–450

84. Pagnanelli F, Moscardini E, Granata G, Cerbelli S, Agosta L, Fieramosca A, Toro L (2014) Acid reducing leaching of cathodic powder from spent lithium ion batteries: glucose oxidative pathways and particle area evolution. J Ind Eng Chem 20(5):3201–3207

85. Chen X, Ma H, Luo C, Zhou T (2017) Recovery of valuable metals from waste cathode materials of spent lithium-ion batteries using mild phosphoric acid. J Hazard Mater 326:77–86

86. Zhang J, Hu J, Liu Y, Jing Q, Yang C, Chen Y, Wang C (2019) Sustainable and facile method for the selective recovery of lithium from cathode scrap of spent $LiFePO_4$ batteries. ACS Sustain Chem Eng 7(6):5626–5631

87. Yang Y, Zheng X, Cao H, Zhao C, Lin X, Ning P, Zhang Y, Jin W, Sun Z (2017) A closed-loop process for selective metal recovery from spent lithium iron phosphate batteries through mechanochemical activation. ACS Sustain Chem Eng 5(11):9972–9980

88. Li H, Xing S, Liu Y, Li F, Guo H, Kuang G (2017) Recovery of lithium, iron, and phosphorus from spent $LiFePO_4$ batteries using stoichiometric sulfuric acid leaching system. ACS Sustain Chem Eng 5(9):8017–8024

89. Huang Y, Han G, Liu J, Chai W, Wang W, Yang S, Su S (2016) A stepwise recovery of metals from hybrid cathodes of spent Li-ion batteries with leaching-flotation-precipitation process. J Power Sources 325:555–564

90. Golmohammadzadeh R, Faraji F, Rashchi F (2018) Recovery of lithium and cobalt from spent lithium ion batteries (LIBs) using organic acids as leaching reagents: a review. Resour Conserv Recycl 136:418–435

91. Arshad F, Li L, Amin K, Fan E, Manurkar N, Ahmad A, Yang J, Wu F, Chen R (2020) A comprehensive review of the advancement in recycling the anode and electrolyte from spent lithium ion batteries. ACS Sustain Chem Eng 8(36):13527–13554

92. Li L, Fan E, Guan Y, Zhang X, Xue Q, Wei L, Wu F, Chen R (2017) Sustainable recovery of cathode materials from spent lithium-ion batteries using lactic acid leaching system. ACS Sustain Chem Eng 5(6):5224–5233

93. Liu X, Ren D, Hsu H, Feng X, Xu G-L, Zhuang M, Gao H, Lu L, Han X, Chu Z (2018) Thermal runaway of lithium-ion batteries without internal short circuit. Joule 2(10):2047–2064

94. Gao W, Song J, Cao H, Lin X, Zhang X, Zheng X, Zhang Y, Sun Z (2018) Selective recovery of valuable metals from spent lithium-ion batteries–process development and kinetics evaluation. J Clean Prod 178:833–845

95. Gao W, Zhang X, Zheng X, Lin X, Cao H, Zhang Y, Sun Z (2017) Lithium carbonate recovery from cathode scrap of spent lithium-ion battery: a closed-loop process. Environ Sci Technol 51(3):1662–1669

96. Li L, Bian Y, Zhang X, Xue Q, Fan E, Wu F, Chen R (2018) Economical recycling process for spent lithium-ion batteries and macro-and micro-scale mechanistic study. J Power Sources 377:70–79

97. Zhang X, Bian Y, Xu S, Fan E, Xue Q, Guan Y, Wu F, Li L, Chen R (2018) Innovative application of acid leaching to regenerate Li $(Ni_1/3Co_1/3Mn_1/3)$ O_2 cathodes from spent lithium-ion batteries. ACS Sustain. Chem. Eng. 6(5):5959–5968

98. He L-P, Sun S-Y, Yu J-G (2018) Performance of LiNi1/3Co1/3Mn1/3O2 prepared from spent lithium-ion batteries by a carbonate co-precipitation method. Ceram Int 44(1):351–357

99. Zhang X, Cao H, Xie Y, Ning P, An H, You H, Nawaz F (2015) A closed-loop process for recycling $LiNi_1/3Co_1/3Mn_1/3O_2$ from the cathode scraps of lithium-ion batteries: Process optimization and kinetics analysis. Sep Purif Technol 150:186–195

100. Pant D, Dolker T (2017) Green and facile method for the recovery of spent lithium nickel manganese cobalt oxide (NMC) based lithium ion batteries. Waste Manage 60:689–695

101. Song D, Wang X, Nie H, Shi H, Wang D, Guo F, Shi X, Zhang L (2014) Heat treatment of $LiCoO_2$ recovered from cathode scraps with solvent method. J Power Sources 249:137–141

102. Nayaka G, Pai K, Santhosh G, Manjanna J (2016) Dissolution of cathode active material of spent Li-ion batteries using tartaric acid and ascorbic acid mixture to recover Co. Hydrometallurgy 161:54–57

103. Zeng X, Li J, Shen B (2015) Novel approach to recover cobalt and lithium from spent lithium-ion battery using oxalic acid. J Hazard Mater 295:112–118

104. Fan E, Li L, Zhang X, Bian Y, Xue Q, Wu J, Wu F, Chen R (2018) Selective recovery of Li and Fe from spent lithium-ion batteries by an environmentally friendly mechanochemical approach. ACS Sustain Chem Eng 6(8):11029–11035

Chapter 7
The Ways Ahead

Although great efforts have been made in developing data science technology for benefitting full-lifespan management of Li-ion batteries, many knowledge gaps still exist. This chapter summarizes these challenges, future trends, and promising solutions to boost the development of data science solutions in the management of battery manufacturing, operation, and reutilization, respectively. This could further inform the selections of data science methodology and academic research agendas alike, thus boosting progress in data science-based battery full-lifespan management on different technology readiness levels.

7.1 Data Science-Based Battery Manufacturing

Efficient data science-based management of battery manufacturing, hampered by the complicated intermediate stages within the battery manufacturing chain, is becoming a key but challenging research direction, as it could play a direct and pivotal role in affecting manufactured battery performance. Considering the huge requirement on battery with high performance for larger-scale energy storage applications, we outline four challenges and ways ahead directions, as illustrated in Fig. 7.1, with the overarching target of achieving more technological innovation and breakthroughs in data science-based management of battery manufacturing.

7.1.1 Continuous Manufacturing Line

The current established battery manufacturing line contains stages with various material flows. For example, coating and calendering are continuous, electrode cutting, electrolyte filling, forming, and testing are discrete, while mixing is batch-wise.

© The Author(s) 2022
K. Liu et al., *Data Science-Based Full-Lifespan Management of Lithium-Ion Battery*, Green Energy and Technology,
https://doi.org/10.1007/978-3-031-01340-9_7

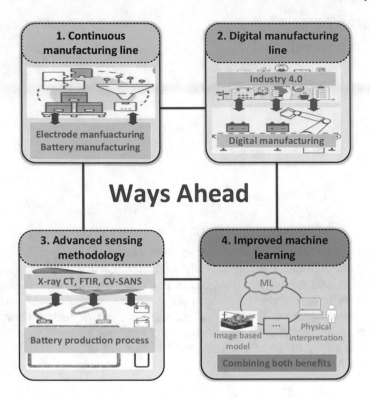

Fig. 7.1 Challenges and ways ahead for data science-based management for better battery manufacturing

Traditionally, as a batch-wise process, slurry mixing would cause variations right from the beginning of electrode manufacturing. A suggested way is to improve mixing technology and make it become a fully continuous process. The obvious benefits of continuous manufacturing line are it requires less manpower to operate but also can include high-precision dosing systems and in-line quality monitoring and analysis. For large-scale battery manufacturing line, achieving continuous manufacturing could also address the bottlenecks in terms of time and cost [1].

7.1.2 Digital Manufacturing Line

Due to the superiorities in terms of predictive maintenance and real-time quality control, Industry 4.0 with digitization is becoming ideal for battery manufacturing [2]. In this case, once the characteristics of the intermediate manufacturing product are measured, the machine corresponding to the next step will adjust its settings for correct further processing. Several measures including equipment automation, digital twin of products and processes have been taken to implement digital manufacturing lines [3], which face their own issues in communication protocols, network security, and financial investment [4]. Therefore, the digital battery manufacturing line becomes a promising direction as the technology cost reduces, the utilization of data science and cloud tool represents less of a barrier. Moreover, with the digitization of the battery manufacturing line, an in-depth understanding of manufacturing equipment, parameters, data collection, and processing is worth being explored.

7.1.3 Advanced Sensing Methodology

Battery manufacturing is a complex process that encompasses multidisciplinary disciplines in material, chemical, and mechanical operations. As the most complex stages, the strongly coupled parameters from coating and drying processes would highly influence electrode and battery performance in terms of material, electrochemical, and mechanical properties. Current researches mostly focus on the off situ electrode characterizations such as surface morphology and element distributions, while in situ methodologies are still poorly adopted. For example, for the drying process, although a few in situ characterization methods have been adopted to explore the drying rate, binder and particle distribution, the dynamic information acquired for drying is still limited [5–7]. To further obtain more qualitative and quantitative data for data science activities, advanced sensing methodologies such as X-ray CT [8], Fourier transform infrared microscopy [9, 10], contrast-variation small-angle neutron scattering [11] are suggested to be adopted.

In the light of this, advanced sensing methodology is crucial for the next generation battery manufacturing management in the medium and long runs. Future ways ahead could focus on in situ sensing methods that could accurately access important information about manufacturing parameters, and deeper insight into the electrochemical measurement of battery manufacturing. Moreover, 3D image sensing methodology with computational models has illustrated substantial benefits in battery properties [12]. This methodology also attempts to give more in situ information on battery manufacturing, further benefitting the generation of new data and profound observation for more efficient data science management of battery manufacturing.

7.1.4 Improved Machine Learning

Data science-based models could bring many merits for parameter analyses in battery manufacturing, such as greatly reducing the expertise required to use model. However, pure data-driven models still present limitations, such as the inability to give physical insights into the battery manufacturing line. This would significantly hinder battery manufacturers to optimize their manufacturing line. In addition, the parameter analysis capabilities of pure data science models are highly influenced by the quality of the data used, because these data must cover enough information to ensure that the model can be well trained. In this context, it is important to explore how to improve the pure data science model by combining it with other powerful tools. One key lies in the design of data physics-driven models by coupling the physical elements of battery manufacturing to machine learning approaches to further help provide physical insights for battery manufacturing parameter analysis. In addition, image-based models that link relevant X-rays and other imaging information to describe battery micro behaviour have also become powerful tools in the battery field [13, 14]. In this context, hybrid tools through combining the benefits of machine learning and image-based model are suggested to improve the interpretability of data science tools and get more valuable information from images for wider battery manufacturing data analyses.

7.2 Data Science-Based Battery Operation

Battery operation management, featured by its multidisciplinary nature, is becoming a fast growing research area, as there are increasingly stringent regulations on battery performance for large-scale transportation electrification applications. Considering potential scientific importance and engineering application requirements, several data science-based research directions and trends in this field, from the perspectives of battery operation modelling and state estimation, lifetime prognostics, fault diagnosis, and battery charging, are outlined, with the overarching target of stimulating more technical innovations and transformative breakthroughs.

7.2.1 Operation Modelling and State Estimation

Although the current data science-based battery operation modelling and state estimation approaches have made great progress, the following challenges still exist and need to be further improved, as illustrated in Fig. 7.2.

Robust and simplified operation modelling: For battery operation modelling, the widely utilized equivalent circuit model (ECM) lacks enough chemical significance, making it become difficult to describe many dynamic characteristics within a battery. Therefore, it is recommended to combine battery electrochemical elements into ECMs. The machine learning-based model is usually trained under a specific condition, further causing generalization issues in real battery operations. This requires the model could be trained under comprehensive operation cases. Therefore, the training data should consider the elements of battery degradation, hysteresis, charging/discharging rates, and temperatures. On the other hand, another big issue for battery operation modelling is the trade-off between computational efforts and

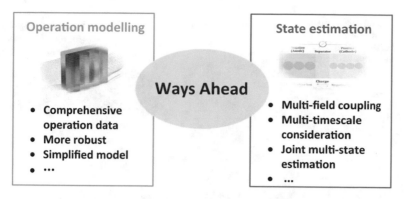

Fig. 7.2 Ways ahead for better data science-based battery operation modelling and state estimation

computing resources. In the light of this, another future research direction should focus on the reduction and simplification of battery operation modelling.

Joint multi-state estimations: There are a number of data science-based battery single state estimation methods reported in the literature, whereas the research of at least two-state joint estimation is still limited. It should be known that battery internal states are actually coupled and interact with each other. Estimating one state independently and ignoring other states can obtain satisfactory results only under certain constraints. In the light of this, according to the multi-field coupling of electro, thermal, ageing, and mechanical conditions of a battery, devising advanced data science-based methods such as fractional-order calculus [14] and multi-time scale estimator [16] to effectively enhance battery multi-state joint estimation performance, with a reliable computing efficiency, becomes another promising research direction.

7.2.2 Lifetime Prognostics

Battery lifetime prognostics is also a key and hot research topic in battery operation management. Although great data science efforts have been made in this field, several challenges from Fig. 7.3 are still existed as:

Battery degradation identification: The pure data science-based solutions, especially just using machine learning technology, are difficult to explain battery degradation mechanisms. It would become meaningful to integrate the information of battery degradation mechanism with the lifetime prognostic approaches. As such, combining machine learning with physical information about battery ageing is a promising research direction. As some battery ageing data curves such as IC/DV contain battery degradation mechanisms, one suggested way is to first collect IC/DV data for uncovering battery ageing mechanisms, then couple this information into

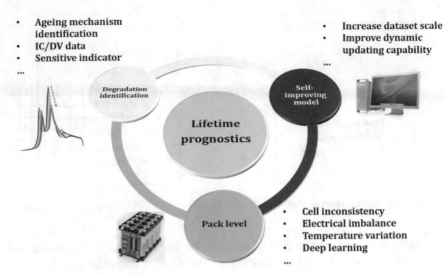

Fig. 7.3 Ways ahead for better data science-based battery lifetime prognostics

machine learning. In this way, IC/DV data could reflect sensitive indicators of battery ageing, further benefitting battery lifetime prognostics.

Self-improving model via online data: Li-ion battery degradation is sensitive to the operation cases. Effectively predicting battery lifetime under conditions different from the training cases is a challenge. The difference between the laboratory cases for model development and the real operation conditions limits the wider applications of data science-based approaches. This could be improved in two ways: (1) the scale of an experimental ageing dataset can be increased to cover more battery degradation information under wider operation cases, further improving the predictability of derived data science-based solutions. However, this would also lead to the increased cost of the battery ageing tests. (2) Improving the dynamic updating capability of data science-based models developed offline is worthy of further research, because it could pave a way to the self-improving model.

Lifetime prognostics at pack level: To date, most of battery lifetime prognostics research is explored under battery cell level. However, in real applications, numerous cells require to be connected with series or parallel forms to construct a battery pack for providing enough energy and power. Understanding battery pack degradation requires knowledge beyond the cell level, considering additional effect elements including cell inconsistency, electrical imbalance and temperature variations among cells. All these issues complicate accurate lifetime prognostics modelling for a battery pack. The advances in the state-of-the-art deep learning tools are foreseen to introduce some ways to these issues. Some deep neural networks such as convolutional neural network and generative adversarial network have the ability for highly complicated nonlinear fitting and become good candidates for handling these issues. The use of such deep learning tools or similar self-learning solutions also becomes a promising way for battery lifetime prognostics at the pack level.

7.2.3   Fault Diagnostics

Battery failure is a very large potential hazard to vehicles, so the battery fault diagnosis is becoming a research hotspot. At the same time, battery failures are concealed and have a long incubation period, which brings great challenges to its diagnosis. The current fault diagnosis is still based on feature detection, and the early warning capability of hidden faults needs to be improved. With the development of sensor and artificial intelligence technologies, future development of battery fault diagnosis has the following trends, as illustrated in Fig. 7.4.

(1)   **Diversification of sensing signals.** The development of smart sensors makes it possible for the smart perception of batteries. Some currently unmeasurable parameters will become measurable through built-in smart sensors, such as the battery internal temperature, pressure or strain inside the battery, internal gas concentration and composition, positive and negative absolute voltages, and impedance spectra. These signals bring important input to the fault diagnosis of the battery, changing the current defect that there is only insufficient information such as battery voltage, current, and temperature. It is worth mentioning that the online measurement technology of EIS provides very valuable information for the battery fault diagnosis, and the development of EIS chips in future is an important direction.

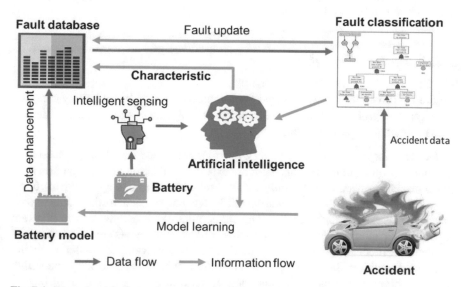

Fig. 7.4 Ways ahead for better data science-based battery fault diagnosis

(2) **High-fidelity battery models**. The current battery model has a contradiction between complexity and accuracy. However, the high-precision state estimation of the battery requires high-precision physical and electrochemical models. Cloud computing makes the application of this highly complex model and the adaptive updating of model parameters based on data become a reality. For the fault diagnosis model, the key technology is how to establish the mapping relationship between the sensor signal and the internal state of the battery from the battery mechanism, and how to establish the mapping relationship between the fault type and the battery model.

(3) **Intelligent diagnosis and decision technologies**. The battery fault characteristics are deeply mined by machine learning. Model, data, and machine learning are becoming the three core elements for battery fault diagnosis. The development of big data has produced a large amount of data. Machine learning technology is an important means and powerful weapon to explore massive data. The model-driven and data-driven fault diagnosis methods urgently need the support of machine learning. The parameters of the physical model need to be learned and improved by machine learning. In addition, how to extract fault features from massive data is a very challenging task. Machine learning can effectively and deeply mine the hidden fault features in the data, which can greatly improve the fault recognition rate and reduce the false alarm rate.

(4) **End-edge-cloud collaboration**. The application of digital twin technology gives a new concept of battery network management and service. The digital twin technology and cloud collaboration of the future battery management system will establish a battery fault diagnosis algorithm based on outlier mining, break through the limitations of computing power and storage space of traditional battery management, and realize the refined safety management of the full battery life cycle.

7.2.4 Battery Charging

Figure 7.5 illustrates the ways ahead for better data science-based battery charging management, which includes three main parts: robustness improvement, thermal management, and pack level charging.

Robustness improvement: For battery charging, as numerous data science-based explorations on charging management are experimental or empirical in nature, their performance has been explored only under a limited range of battery chemistry, operation factors and cases. These results are difficult to be extended to other battery types or conditions, as supported by the frequent conflicting observations from different reports. Besides, lots of model-based optimal charging strategies are based on equivalent circuit model or single particle model, and only be validated under a specific operating condition. Such strategies would become inaccurate especially under high power or high current cases. In this context, more in-depth exploration of battery charging behaviours and robust data science-based battery models are required for

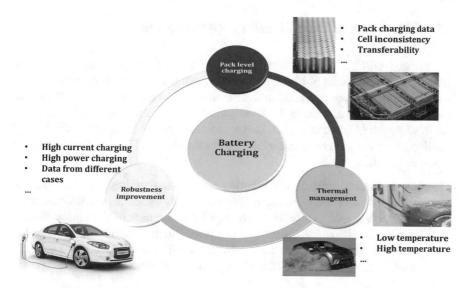

Fig. 7.5 Ways ahead for better data science-based battery charging management

efficient battery charging strategy design that could generalize well under different operational conditions.

Thermal management under low/high temperatures: Thermal management is a key element in battery charging. To date, existing charging strategies are mainly designed under normal temperature conditions without considering the cases of low or high temperature charging. Due to the increased number of electrical vehicles is deployed in both colder and hotter climates, battery charging strategies under such conditions become increasingly critical. In this context, experimental data under extreme charging conditions are required to design suitable data science-based charging strategies. Then these charging strategies could equip with suitable thermal management solutions to preheat or cool batteries. This could significantly improve battery safety during the charging period and performance such as battery lifetime.

Pack level charging: To further benefit the charging performance of each individual cell within a pack, charging strategies considering the cell inconsistency are required to avoid local overcharge or safety issues. Currently, multi-objective charging strategies have been well designed for single battery cells, but their transferability, influence and costs in battery packs have not been fully explored. Charging strategies with the ability to improve the performance of a single cell would also lead to uneven currents or temperatures when executed on the battery pack. As fast charging would amplify heterogeneity, such charging research considering the effects of cell-to-cell difference is urgently required. In this context, experimental data of different cells within battery pack charging are required. Then the data science-based battery pack charging solutions could be explored to integrate cell and pack management, further improving charging performance at the pack level.

7.3 Data Science-Based Battery Reutilization

The echelon utilization and material recycling are the key links to building the closed-loop management of Li-ion batteries in the full life cycle. They have obvious significance in resource recycling and environmental protection. With the explosive development of EVs and the rapid development of the Internet, big data, and artificial intelligence (AI) technologies, battery management is entering the digital and intelligent stage in the entire life cycle. As summarized in Fig. 7.6, there are three important trends.

(1) **New industrial structure is forming and reconstructing**. Massive zero/low marginal cost Li-ion batteries in EVs will greatly enrich the flexible resources of the power grid and promote the rapid development of the mobile energy Internet. In 2030, EV ownership in China will reach 100 million, and the power of on-board power Li-ion batteries will exceed 1 billion kilowatts, which is equivalent to 50 Three Gorges power stations. The on-board power

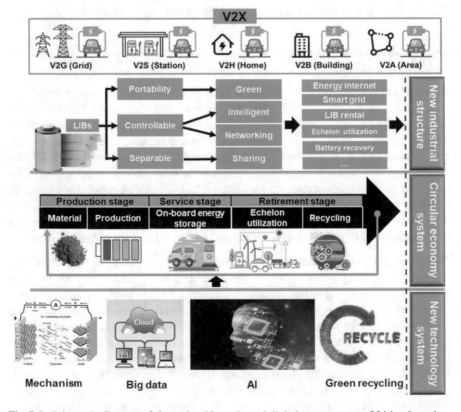

Fig. 7.6 Schematic diagram of the entire life cycle and digital management of Li-ion batteries, reprinted from [15], with permission from Elsevier

batteries of mass EVs can absorb the excess power in the power grid. What we need to do is to make these EVs charge and discharge at the right time. The power battery will make the energy Internet connect more widely, flexibly, stably, and strengthen the cooperation with other elements. Under the framework of energy Internet, smart grid, battery leasing, battery recycling, echelon utilization, sharing economy, and other new industrial structures and modes are forming and developing.

(2) **Circular economy is further developing and deepening**. The entire life cycle of Li-ion batteries can be divided into production, service, and retirement stages. Li-ion batteries provide power for EVs during service, which is also an important part of the energy Internet. When the battery is retired from the EV and enters the stage of echelon utilization, it can directly participate in the energy Internet. When Li-ion battery cannot meet the requirements of echelon utilization scenarios, it will enter the recycling link. Recycling valuable metal of the spent battery has great environmental protection and economic value. The recycled battery materials can be reused for battery remanufacturing, which is of great significance to alleviate the resource crisis. By constructing the cycle economy model of the full life cycle of Li-ion batteries, the value of Li-ion batteries can be maximized, which is of great significance to the sustainable development of Li-ion batteries.

(3) **The new technologies are applied in the full life cycle of Li-ion batteries**. First, the ageing and failure mechanism of Li-ion batteries will be clearer, which is very critical for the fine management of Li-ion batteries. Second, the high-quality operation data in the entire life cycle of Li-ion batteries is very valuable. These data are the basis of accurate state estimation and safety management of Li-ion batteries, and they also facilitate the quick evaluation of residual value of the retired Li-ion batteries. In recent years, the intelligent BMS integrating mechanism model, cloud data and artificial intelligence technologies have become a hot research topic, which promotes the accurate and efficient management, optimization and control of Li-ion batteries in the full life cycle. For Li-ion batteries recycling, the development of new recycling methods with characteristics of green, highly efficient, low energy consumption, short process is an important research field. We think that the following three aspects are favourable trends: (a) the study of selective leaching separation of Li will greatly improve the economic benefits of recycling; (b) the development of co-extractant for the extraction of Ni, Co, and Mn metals will shorten the separation process and reduce costs; (c) direct electrode recycling technology is an important development direction.

(4) **Sustainable and green development have attracted more and more attention**. Under the requirements of carbon neutralization, the carbon emission in battery recycling and remanufacturing needs to be paid special attention. Battery recycling has high economic value, but its impact on the environment is a topic that needs to be expounded. As a powerful evaluation tool of environment, resources, and cost, lifecycle assessment (LCA) has received great attention in the full lifecycle management of Li-ion batteries in recent years.

The environmental and resource burdens are evaluated by the LCA method in the full life cycle of Li-ion batteries is of great significance to improve process flow and products, risk assessment and policy decision-making.

7.4 Summary

Li-ion battery represents one of the promising energy storage solutions for many applications such as transport electrification and smart grid, owing to its high energy density, reliable service life. Battery management technologies are developing rapidly in the international market and are also a hot research topic. However, there are many technical challenges in battery full-lifespan applications, demanding state-of-the-art data science approaches. Firstly, battery properties such as cost, reliability, energy density, and life are directly determined by its manufacturing process. It is thus vital to develop suitable solutions for understanding and analysing battery intermediate manufacturing processes in the pursuit of smarter battery manufacturing management. Besides, owing to complicated electrochemical dynamics, numerous tasks including battery operation modelling, state estimation, lifetime prognostics, fault diagnosis, and charging must be done to well and efficiently manage battery during its operation stage. Moreover, to make full use of battery residual value, the retired battery will be reutilized in second-life applications such as grid energy storage. To further comply with environmental and health benefits, batteries need to be recycled finally when they get either spoilt or non-functional. With the rapid development of AI and machine learning technologies, data science-based applications have drawn much attention and become a research hotspot in the field of battery full-lifespan management. After well-designing proper data science solutions, significant enhancement can be achieved for more effective battery management from the aforementioned three parts. However, as relative new and prospective research, currently there is no book to systematically introduce and describe the battery full-lifespan management particular from data science application perspective to our best knowledge.

In this book, data science-based battery full-lifespan management strategies are comprehensively reviewed and discussed. In Chaps. 1 and 2, i.e. (1) the introduction to Li-ion battery and related management, (2) the key stages of battery full-lifespan and the basics of data science technology, give illustrative descriptions on the research focus. Then in Chaps. 3–6, the new and emerging data science technologies for full-lifespan management of Li-ion battery from three key aspects, i.e. (1) battery manufacturing management, (2) battery operation management, and (3) battery reutilization management, are discussed with plentiful case studies. Finally, this chapter overviews the key challenges, future roadmap in all these three parts.

In a nutshell, to make full use of battery to support power/energy and ensure its safety, efficiency and performance, reliable data science methods are required in the battery full-lifespan management, but many corresponding technologies are immature. None of the data science solutions can be regarded as a one-size-fits-all solution; instead, there are inherent trade-offs between complexity and performance in different applications. To widen the data science applications for battery management, three key features are given in this book: (1) the concept of full-lifespan management of Li-ion battery is proposed and the state-of-the-art data science technologies to handle related key tasks are described. (2) Case studies of deriving various data science technologies to benefit battery manufacturing, operation and reutilization are systematically introduced, which proves that data science is a promising route to improve the full-lifespan management of battery. (3) Valuable guidance for the challenges, future trends, and promising solutions to benefit data science-based battery full-lifespan management are provided. With the above arrangement, we hope that this book provides useful reference points to support the design of data science-based battery management solutions during its lifespan, while a brand-new hologram to make full use of battery during full-lifespan will be formulated, further boosting the advancement of AI and low-carbon technologies.

References

1. Wood III DL, Li J, Daniel C (2015) Prospects for reducing the processing cost of lithium ion batteries. J Power Sources 275:234–242
2. Chen B, Wan J, Shu L, Li P, Mukherjee M, Yin B (2017) Smart factory of Industry 4.0: key technologies, application case, and challenges. IEEE Access 6:6505–6519
3. Kozák Š, Ružický E, Štefanovič J, Schindler F (2018) Research and education for Industry 4.0: present development. In: Proceedings of 2018 cybernetics & informatics (K&I), pp 1–8
4. Vaidya S, Ambad P, Bhosle S (2018) Industry 4.0—a glimpse. Procedia Manuf 20:233–238
5. Finegan DP, Scheel M, Robinson JB, Tjaden B, Hunt I, Mason TJ, Millichamp J, Di Michiel M, Offer GJ, Hinds G (2015) In-operando high-speed tomography of lithium-ion batteries during thermal runaway. Nat Commun 6(1):1–10
6. Hsieh A, Bhadra S, Hertzberg B, Gjeltema P, Goy A, Fleischer JW, Steingart DA (2015) Electrochemical-acoustic time of flight: in operando correlation of physical dynamics with battery charge and health. Energy Environ Sci 8(5):1569–1577
7. Lim S, Ahn KH, Yamamura M (2013) Latex migration in battery slurries during drying. Langmuir 29(26):8233–8244
8. Higa K, Zhao H, Parkinson DY, Barnard H, Ling M, Liu G, Srinivasan V (2017) Electrode slurry particle density mapping using X-ray radiography. J Electrochem Soc 164(2):A380
9. Ni K, Wang X, Tao Z, Yang J, Shu N, Ye J, Pan F, Xie J, Tan Z, Sun X (2019) In operando probing of lithium-ion storage on single-layer graphene. Adv Mater 31(23):1808091
10. Pérez-Villar S, Lanz P, Schneider H, Novák P (2013) Characterization of a model solid electrolyte interphase/carbon interface by combined in situ Raman/Fourier transform infrared microscopy. Electrochim Acta 106:506–515
11. Kusano T, Hiroi T, Amemiya K, Ando M, Takahashi T, Shibayama M (2015) Structural evolution of a catalyst ink for fuel cells during the drying process investigated by CV-SANS. Eur Polym J 47(8):546–555
12. Jiang Z, Li J, Yang Y, Mu L, Wei C, Yu X, Pianetta P, Zhao K, Cloetens P, Lin F (2020) Machine-learning-revealed statistics of the particle-carbon/binder detachment in lithium-ion battery cathodes. Nat Commun 11(1):1–9
13. Usseglio-Viretta FL, Colclasure A, Mistry AN, Claver KPY, Pouraghajan F, Finegan DP, Heenan TM, Abraham D, Mukherjee PP, Wheeler D (2018) Resolving the discrepancy in

tortuosity factor estimation for Li-ion battery electrodes through micro-macro modeling and experiment. J Electrochem Soc 165(14):A3403

14. Badmos O, Kopp A, Bernthaler T, Schneider G (2020) Image-based defect detection in lithium-ion battery electrode using convolutional neural networks. J Manuf Syst 31(4):885–897

15. Lai X, Huang Y, Gu H, Deng C, Han X, Feng X, Zheng Y (2021) Turning waste into wealth: a systematic review on echelon utilization and material recovery of the retired lithium-ion batteries. Energy Storage Mater 40:96–123

16. Xitian He, Bingxiang Sun, Weige Zhang, Xinyuan Fan, Xiaojia Su, and Haijun Ruan (2022) "Multi-time scale variable-order equivalent circuit model for virtual battery considering initial polarization condition of lithium-ion battery". Energy : 123084

Printed in the United States
by Baker & Taylor Publisher Services